Soils, Rocks, and Landforms

INVESTIGATIONS GUIDE

FOSS PATHWAYS™
Developed at
The Lawrence Hall of Science

Published and Distributed by
Delta Education/School Specialty

FOSS Staff at the Lawrence Hall of Science, University of California, Berkeley
Codirectors and Developers: Larry Malone and Linda De Lucchi
Curriculum Coordinator and Developer: Jessica Penchos
Technology Development Manager: Susan Ketchner
Video Developer and Project Management: David Lippman
Curriculum Developers: Brian Campbell, Kimi Hosoume, Erica Beck Spencer, Joanna Totino, Diana Velez, Natalie Yakushiji
Assessment Developers: Arthur Camins, Kathy Long, Eileen Patrick
FOSSweb Development: Kate Jordan, Frank Kusiak, Shan Tsai
Design and Illustrations: Carol Sevilla, Susan Stanley

School Specialty Team
Bonnie A. Piotrowski, Senior Director, Product Development
Project Team: Sandra Burke, Joanne Deegan, Debra Housel, Jennifer McKenna, Angela Miccinello, Malgorzata Richard

Cover Series and Design
3CD Chicago

Thank you to all FOSS Grades K-5 Trial Teachers
Heather Ballard, Wilson Elementary, Coppell, TX; Mirith Ballestas De Barroso, Treasure Forest Elementary, Houston, TX; Terra L. Barton, Harry McKillop Elementary, Melissa, TX; Rhonda Bernard, Frances E. Norton Elementary, Allen, TX; Theresa Bissonnette, East Millbrook Magnet Middle School, Raleigh, NC; Peter Blackstone, Hall Elementary School, Portland, ME; Tiffani Brisco, Seven Hills Elementary, Newark, TX; Darrow Brown, Lake Myra Elementary School, Wendell, NC; Heather Callaghan, Olive Chapel Elementary, Apex, NC; Katie Cannon, Las Colinas Elementary, Irving, TX; Elaine M. Cansler, Brassfi eld Road Elementary School, Raleigh, NC; Kristy Cash, Wilson Elementary, Coppell, TX; Monica Coles, Swift Creek Elementary School, Raleigh, NC; Shirley Conner, Ocean Avenue Elementary School, Portland, ME; Sally Connolly, Cape Elizabeth Middle School, Cape Elizabeth, ME; Melissa Cook-Airhart, Harry McKillop Elementary, Melissa, TX; Melissa Costa, Olive Chapel Elementary, Apex, NC; Hillary P. Croissant, Harry McKillop Elementary, Melissa, TX; Rene Custeau, Hall Elementary School, Portland, ME; Nancy Davis, Martha and Josh Morriss Mathematics and Engineering Elementary School, Texarkana, TX; Nancy Deveneau, Wilson Elementary, Coppell, TX; Karen Diaz, Las Colinas Elementary, Irving, TX; Marlana Dumas, Las Colinas Elementary, Irving, TX; Mary Evans, R.E. Good Elementary School, Carrollton, TX; Jacquelyn Farley, Moss Haven Elementary, Dallas, TX; Corinna Ferrier, Oak Forest Elementary, Humble, TX; Allison Fike, Wilson Elementary, Coppell, TX; Barbara Fugitt, Martha and Josh Morriss Mathematics and Engineering Elementary School, Texarkana, TX; Colleen Garvey, Farmington Woods Elementary, Cary, NC; Judy Geller, Bentley Elementary School, Oakland, CA; Erin Gibson, Las Colinas Elementary, Irving, TX; Kelli Gobel, Melissa Ridge Intermediate School, Melissa, TX; Dollie Green, Melissa Ridge Intermediate School, Melissa, TX; Brenda Lee Harrigan, Bentley Elementary School, Oakland, CA; Cori Harris, Samuel Beck Elementary, Trophy Club, TX; Kim Hayes, Martha and Josh Morriss Mathematics and Engineering Elementary School, Texarkana, TX; Staci Lynn Hester, Lacy Elementary School, Raleigh, NC; Amanda Hill, Las Colinas Elementary, Irving, TX; Margaret Hillman, Ocean Avenue Elementary School, Portland, ME; Cindy Holder, Oak Forest Elementary, Humble, TX; Sarah Huber, Hodge Road Elementary, Knightdale, NC; Susan Jacobs, Granger Elementary, Keller, TX; Carol Kellum, Wallace Elementary, Dallas, TX; Jennifer A. Kelly, Hall Elementary School, Portland, ME; Brittani Kern, Fox Road Elementary, Raleigh, NC; Jodi Lay, Lufkin Road Middle School, Apex, NC; Melissa Lourenco, Lake Myra Elementary School, Wendell, NC; Ana Martinez, RISD Academy, Dallas, TX; Shaheen Mavani, Las Colinas Elementary, Irving, TX; Mary Linley McClendon, Math Science Technology Magnet School, Richardson, TX; Adam McKay, Davis Drive Elementary, Cary, NC; Leslie Meadows, Lake Myra Elementary School, Wendell, NC; Anne Mechler, J. Erik Jonsson Community School, Dallas, TX; Anne Miller, J. Erik Jonsson Community School, Dallas, TX; Shirley Diann Miller, The Rice School, Houston, TX; Keri Minier, Las Colinas Elementary, Irving, TX; Stephanie Renee Nance, T.H. Rogers Elementary, Houston, TX; Cynthia Nilsen, Peaks Island School, Peaks Island, ME; Elizabeth Noble, Las Colinas Elementary, Irving, TX; Courtney Noonan, Shadow Oaks Elementary School, Houston, TX; Sarah Peden, Aversboro Elementary School, Garner, NC; Carrie Prince, School at St. George Place, Houston, TX; Marlaina Pritchard, Melissa Ridge Intermediate School, Melissa, TX; Alice Pujol, J. Erik Jonsson Community School, Dallas, TX; Claire Ramsbotham, Cape Elizabeth Middle School, Cape Elizabeth, ME; Paul Rendon, Bentley Elementary, Oakland, CA; Janette Ridley, W.H. Wilson Elementary School, Coppell, TX; Kristina (Crickett) Roberts, W.H. Wilson Elementary School, Coppell, TX; Heather Rogers, Wendell Creative Arts & Science Magnet Elementary School, Wendell, NC; Alissa Royal, Melissa Ridge Intermediate School, Melissa, TX; Megan Runion, Olive Chapel Elementary, Apex, NC; Christy Scheef, J. Erik Jonsson Community School, Dallas, TX; Samrawit Shawl, T.H. Rogers School, Houston, TX; Nicole Spivey, Lake Myra Elementary School, Wendell, NC; Ashley Stephenson, J. Erik Jonsson Community School, Dallas, TX; Jolanta Stern, Browning Elementary School, Houston, TX; Gale Stimson, Bentley Elementary, Oakland, CA; Ted Stoeckley, Hall Middle School, Larkspur, CA; Cathryn Sutton, Wilson Elementary, Coppell, TX; Camille Swander, Ocean Avenue Elementary School, Portland, ME; Brandi Swann, Westlawn Elementary School, Texarkana, TX; Robin Taylor, Arapaho Classical Magnet, Richardson, TX; Michael C. Thomas, Forest Lane Academy, Dallas, TX; Jomarga Thompkins, Lockhart Elementary, Houston, TX; Mary Timar, Madera Elementary, Lake Forest, CA; Helena Tongkeamha, White Rock Elementary, Dallas, TX; Linda Trampe, J. Erik Jonsson Community School, Dallas, TX; Charity VanHorn, Fred A. Olds Elementary, Raleigh, NC; Kathleen VanKeuren, Lufkin Road Middle School, Apex, NC; Valerie Vassar, Hall Elementary School, Portland, ME; Megan Veron, Westwood Elementary School, Houston, TX; Mary Margaret Waters, Frances E. Norton Elementary, Allen, TX; Stephanie Robledo Watson, Ridgecrest Elementary School, Houston, TX; Lisa Webb, Madisonville Intermediate, Madisonville, TX; Matt Whaley, Cape Elizabeth Middle School, Cape Elizabeth, ME; Nancy White, Canyon Creek Elementary, Austin, TX; Barbara Yurick, Oak Forest Elementary, Humble, TX; Linda Zittel, Mira Vista Elementary, Richmond, CA

Published and Distributed by Delta Education, a member of the School Specialty Family
The FOSS program was developed in part with the support of the National Science Foundation grant nos. MDR-8751727 and MDR-9150097. However, any opinions, findings, conclusions, statements, and recommendations expressed herein are those of the authors and do not necessarily reflect the views of NSF. FOSSmap was developed in collaboration between the BEAR Center at UC Berkeley and FOSS at the Lawrence Hall of Science.

Soils, Rocks, and Landforms Investigations Guide, 2088725
978-1-64011-850-8
Printing 2 — 1/2024
Bradford & Bigelow, Newburyport, MA

INVESTIGATIONS GUIDE

Table of Contents

Welcome!

FOSS® Pathways is an active learning PreK–5 science program developed at the Lawrence Hall of Science for the Next Generation Science Standards (NGSS). This document will introduce you to the major components of the program to get you started with FOSS.

MEET YOUR MODULE

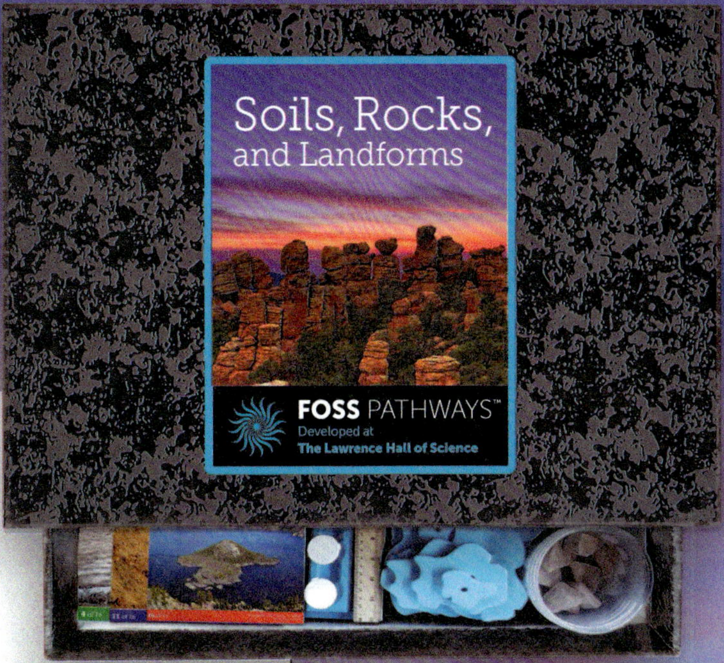

The Soils, Rocks, and Landforms module has three drawers and one plastic box for earth material.

Your FOSS module equipment is organized in one or more large sturdy boxes, called drawers, and two smaller boxes for the print teacher guide and student books. Each drawer has a label on the front listing its contents. Your packing list is always in Drawer 1.

Equipment

FOSS provides the equipment needed for all the investigations, including metric measuring tools. Our high-quality, classroom-tested materials are long-lasting and packaged by investigation to facilitate preparation and clean up. There is enough permanent equipment in each kit for 32 students. Consumable materials are supplied for three uses. Convenient grade-level and refill kits are available.

Getting Started

Teacher Guide

The *Investigations Guide* is a spiral-bound guide containing everything you need to teach the module. FOSS active investigation lesson plans include:

- Three-dimensional learning objectives
- Relevant and local phenomena storylines with driving questions
- Sense-making discussions
- Embedded assessment and "What to Look For" guidance
- Vocabulary reviews
- English language support strategies
- ELA strategies and connections

Side Trips (optional activities outside of the main pathway) provide ways to customize the curriculum instruction by selecting additional pathways, such as environmental literacy, engineering, or a specific practice, such as modeling. Side Trips can be used with the whole class, as student choice challenges, or for differentiation.

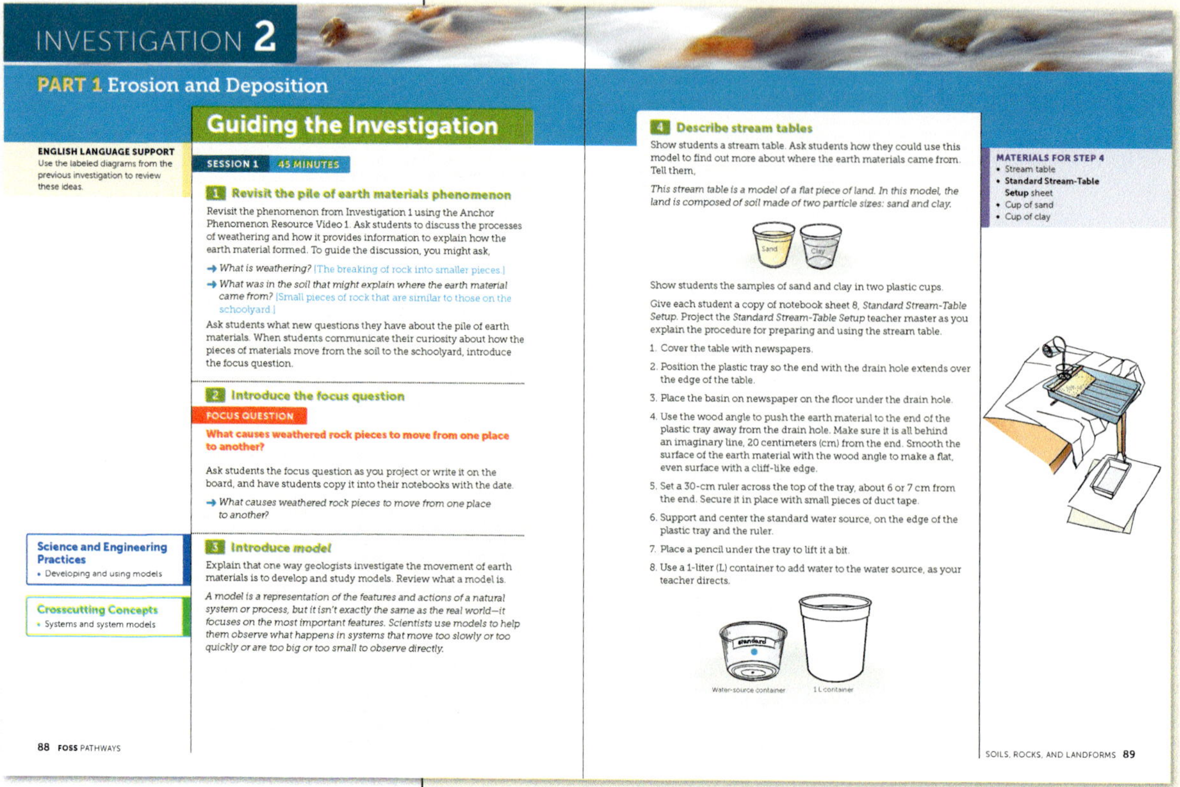

INVESTIGATION 2

PART 1 Erosion and Deposition

Guiding the Investigation

ENGLISH LANGUAGE SUPPORT
Use the labeled diagrams from the previous investigation to review these ideas.

SESSION 1 45 MINUTES

1 Revisit the pile of earth materials phenomenon

Revisit the phenomenon from Investigation 1 using the Anchor Phenomenon Resource Video 1. Ask students to discuss the processes of weathering and how it provides information to explain how the earth material formed. To guide the discussion, you might ask,

→ *What is weathering?* [The breaking of rock into smaller pieces.]

→ *What was in the soil that might explain where the earth material came from?* [Small pieces of rock that are similar to those on the schoolyard.]

Ask students what new questions they have about the pile of earth materials. When students communicate their curiosity about how the pieces of materials move from the soil to the schoolyard, introduce the focus question.

2 Introduce the focus question

FOCUS QUESTION

What causes weathered rock pieces to move from one place to another?

Ask students the focus question as you project or write it on the board, and have students copy it into their notebooks with the date.

→ *What causes weathered rock pieces to move from one place to another?*

Science and Engineering Practices
- Developing and using models

Crosscutting Concepts
- Systems and system models

3 Introduce *model*

Explain that one way geologists investigate the movement of earth materials is to develop and study models. Review what a model is.

A model is a representation of the features and actions of a natural system or process, but it isn't exactly the same as the real world—it focuses on the most important features. Scientists use models to help them observe what happens in systems that move too slowly or too quickly or are too big or too small to observe directly.

88 FOSS PATHWAYS

4 Describe stream tables

Show students a stream table. Ask students how they could use this model to find out more about where the earth materials came from. Tell them,

This stream table is a model of a flat piece of land. In this model, the land is composed of soil made of two particle sizes: sand and clay.

Show students the samples of sand and clay in two plastic cups.

Give each student a copy of notebook sheet 8, *Standard Stream-Table Setup*. Project the *Standard Stream-Table Setup* teacher master as you explain the procedure for preparing and using the stream table.

1. Cover the table with newspapers.
2. Position the plastic tray so the end with the drain hole extends over the edge of the table.
3. Place the basin on newspaper on the floor under the drain hole.
4. Use the wood angle to push the earth material to the end of the plastic tray away from the drain hole. Make sure it is all behind an imaginary line, 20 centimeters (cm) from the end. Smooth the surface of the earth material with the wood angle to make a flat, even surface with a cliff-like edge.
5. Set a 30-cm ruler across the top of the tray, about 6 or 7 cm from the end. Secure it in place with small pieces of duct tape.
6. Support and center the standard water source, on the edge of the plastic tray and the ruler.
7. Place a pencil under the tray to lift it a bit.
8. Use a 1-liter (L) container to add water to the water source, as your teacher directs.

MATERIALS FOR STEP 4
- Stream table
- **Standard Stream-Table Setup** sheet
- Cup of sand
- Cup of clay

SOILS, ROCKS, AND LANDFORMS 89

Student Books

The *FOSS Science Resources* student book contains readings developed to reinforce, extend, or apply core ideas covered during FOSS active investigations. Readings give students opportunities to:

- Use text to obtain, evaluate, and communicate information
- Use evidence to support their ideas during sense-making discussions and focus question responses
- Integrate information from multiple sources
- Interpret graphs, diagrams, and photographs to build understanding

Available in print and as an interactive eBook in English and Spanish.

FOSSILS
TELL A STORY

Ammonites died out about 66 million years ago, but their fossils have been found.

Earth scientists called geologists and **paleontologists** ask questions about the history of Earth. Geologists focus their studies on the structures of rocks and how they formed. Paleontologists ask questions about what life was like millions of years ago. These scientists can't travel back in time, so they look for answers to their questions in rocks. From the **evidence** they find, paleontologists build models to explain how plants and animals lived and what their world was like.

76

Scientists dig through layers of soil and rock.

Digging through the Layers

Over billions of years, Earth's surface has changed dramatically. Areas once covered by water are now dry. Continents have moved. Large areas of Earth's surface were once covered by thick sheets of ice, which have now melted.

Through all these changes, layers of new rock have formed on Earth. Some layers were formed by molten lava that erupted from volcanoes, covered Earth's surface, cooled, and turned into solid rock. Other layers of rock formed from tiny particles of rock called sediment that settled in the ocean, lakes, and swamps. After millions of years these layers of sediment changed into sedimentary rock. By digging through layers of sedimentary rock, earth scientists have determined the age of Earth and discovered the **organisms** that lived at the time the layers of sediment were deposited.

77

The Fossil Record

All the fossils on Earth make up the **fossil record**. It is a valuable source of information about the history of life on Earth.

The fossil record provides evidence of the environment in which ancient organisms lived. An important part of an organism's environment is climate, or the average weather in an area.

We have learned from fossils that Earth's climates have changed over time. For example, fossils of plants that would have thrived in hot and humid areas (tropical jungle) have been found in areas that are now deserts. Such fossil evidence suggests that these areas were wetter millions of years ago than they are today.

Fossilized tree trunks can provide similar kinds of evidence. Trees do not grow as fast when the temperature is cold and dry. A petrified tree trunk with narrow growth rings suggests that the climate was probably cool and dry when the tree was living. A petrified tree trunk with wide growth rings suggests that the climate was probably warm and moist during the tree's lifetime.

Petrified wood is millions of years old and each piece is one of a kind.

This scientist is carefully brushing away the sand on this fossil.

Question for Reflection

- Sometimes scientists find fossils of ocean animals high up on a mountain. What do you think they conclude about the history of that area? Why?

80

81

Students can demonstrate their understanding by answering the "Questions for Reflection" at the end of each in-class reading.

SCAN HERE FOR A TOUR OF FOSSWEB!

FOSSweb

Technology for Learning Anywhere

FOSSweb digital resources are delivered on School Specialty's curriculum platform called ThinkLink.

- Supports single sign-on and class management with Google classroom and learning management systems.
- Provides access to both teacher and student digital resources, including duplication masters, teaching slides, FOSSmap online assessment, streaming videos, and online activities.

FOSSmap Online Assessment

Students in grades 3–5 can take summative assessments online with automatic coding of most responses. Student- and class-level reports help you identify the need for instructional next steps.

Streaming Videos

New engaging content videos in English and Spanish developed to specifically support FOSS investigations.

Online Activities

Engaging, interactive virtual investigations and tutorials that offer additional content support for students.

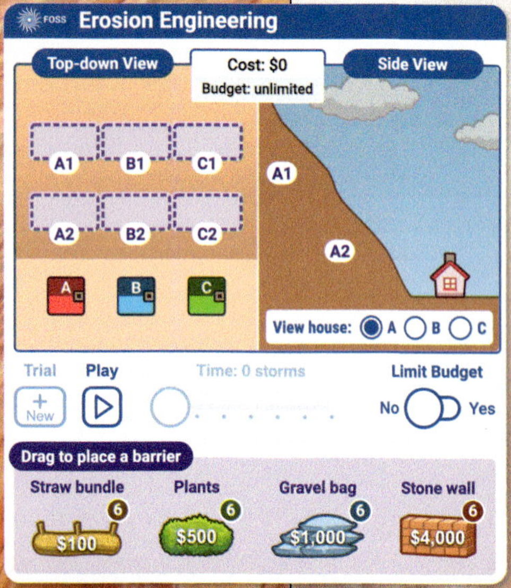

Soils, Rocks, and Landforms

Soils, Rocks, and Landforms

Introduction

The **Soils, Rocks, and Landforms Module** provides students with firsthand experiences with soils and rocks and modeling experiences using tools such as topographic maps and stream tables to engage with the surface of Earth's landscape—the shape and the composition of landforms and the stories they tell of past events.

Student engagement with these ideas is driven by four anchor phenomena that drive students to investigate weathering by water, ice, wind, living organisms; breaking rocks into smaller pieces; erosion or movement of earth materials to new locations; and deposition, the result of that transport process that builds new land. They investigate these phenomena:

- Anchor phenomenon 1—pile of earth material on the schoolyard
- Anchor phenomenon 2—impact to Mount St. Helens
- Anchor phenomenon 3—earth-shaking events
- Anchor phenomenon 4—rock layers, flat and tilted

Students plan and carry out investigations by incrementally changing specific environmental conditions to determine the impact of changing the variables of slope and amount of water in stream tables. Students analyze and interpret data from diagrams and visual representations to build explanations from evidence and make predictions of future events. They develop model mountains and represent the landforms from different perspectives to look for change. Students gain experiences that will contribute to the understanding of the crosscutting concepts of patterns; cause and effect; scale, proportion, and quantity; systems and system models; and stability and change.

CONTENTS

The NGSS Performance Expectations bundled in this module include:

Earth and Space Sciences
4-ESS1-1
4-ESS2-1
4-ESS2-2
4-ESS3-2

Engineering, Technology, and Applications of Science
3–5 ETS1-2

NOTE

The three modules for grade 4 in FOSS Pathways are:

- Energy
- Soils, Rocks, and Landforms
- Senses and Survival

OVERVIEW

Module Matrix
At a Glance

Phenomenon and Storyline	Driving Question and Focus Questions
INV. 1 Soils and Weathering **Phenomenon 1—Pile of earth material on the schoolyard:** On a walk around the schoolyard, students discover some changes at the edge of the playground. They observe a new pile of sand and "dirt" near the bottom of the small hill. **Storyline:** Students plan and carry out investigations comparing properties of four different soils. They learn that soils are composed of essentially the same types of materials (inorganic earth materials and humus), but the amounts of the materials vary. They begin to explore how rocks break into smaller pieces through physical and chemical weathering. Students apply their understanding of the components of soil to the pile of earth materials on the schoolyard.	*What caused this pile of earth material to form?* **FOCUS QUESTIONS:** **What evidence does the composition of soil provide about where it came from?** **What causes big rocks to break into smaller rocks?**
INV. 2 Landforms **Revisit Phenomenon 1—Pile of earth material on the schoolyard:** On a walk around the schoolyard, students discover some changes at the edge of the playground. They observe a new pile of sand and "dirt" near the bottom of the small hill. **Storyline:** Students analyze and interpret the results of erosion and deposition of weathered earth material by flowing water. They use stream-table models to observe that water moves earth materials from one location to another. They investigate the variables of slope and water quantity and plan and conduct their own stream-table investigations. Their findings help students figure out how the earth material was carried to the location on the schoolyard. Students use an online activity, "Erosion Engineering," to collect data on the impact of landslides on human-made structures and compare the effect of four kinds of erosion barriers to reduce the impact.	*What caused this pile of earth material to form?* **FOCUS QUESTIONS:** **What causes weathered rock pieces to move from one place to another?** **How does slope affect erosion and deposition? How do floods affect erosion and deposition?** **What solutions can we design to reduce the impact of erosion and deposition on human structures?**

Content and Disciplinary Core Ideas	Practices and Crosscutting Concepts	NGSS PEs
ESS2.A: Earth materials and systems **ESS2.E:** Biogeology • Soils can be described by their properties. • Soils are composed of different kinds and amounts of earth materials and humus. • Living things affect the characteristics of soil. • The composition of soil provides evidence of its origins. • Weathering is the breakdown of rocks and minerals at or near Earth's surface. • The physical-weathering processes of abrasion, freezing and thawing, and the pressure of tree roots (living things) break rocks and minerals into smaller pieces.	**Science and Engineering Practices** Asking questions Developing and using models Planning and carrying out investigations Analyzing and interpreting data Constructing explanations Engaging in argument from evidence Obtaining, evaluating, and communicating information **Crosscutting Concepts** Patterns Cause and effect Systems and system models	**4-ESS2-1:** Make observations and/or measurements to provide evidence of the effects of weathering or the rate of erosion by water, ice, wind, or vegetation.
ESS2.A: Earth materials and systems **ESS2.E:** Biogeology **ESS3.B:** Natural hazards **ETS1.B:** Developing possible solutions • Weathered rock material can be reshaped into new landforms by the processes of erosion and deposition. • Erosion is the transport (movement) of weathered rock material (sediment) by moving water or wind. • Deposition is the settling of sediment when the speed of moving water or wind declines. • The rate and volume of erosion relate directly to the energy of moving water or wind. • The energy of moving water depends on the mass of water in motion and its velocity. The greater the mass and velocity, the greater the energy. • Scale models can be used to gather data and make decisions about natural systems.	**Science and Engineering Practices** Asking questions and defining problems Developing and using models Planning and carrying out investigations Analyzing and interpreting data Constructing explanations and designing solutions Engaging in argument from evidence Obtaining, evaluating, and communicating information **Crosscutting Concepts** Cause and effect Scale, proportion, and quantity Systems and system models Stability and change Energy and matter	**4-ESS2-1:** Make observations and/or measurements to provide evidence of the effects of weathering or the rate of erosion by water, ice, wind, or vegetation. **4-ESS3-2:** Generate and compare multiple solutions to reduce the impacts of natural Earth processes on humans. **3–5-ETS1-2:** Generate and compare multiple solutions to a problem based on how well each is likely to meet the criteria and constraints of the problem.

OVERVIEW

Module Matrix

At a Glance CONTINUED

Phenomenon and Storyline	Driving Question and Focus Questions

INV. 3 Mapping Earth's Surface

Problem—Hiking to the peak of a mountain: A group of hikers have to solve a problem. They want to hike to the top of a volcanic peak and have selected a mountain in the Cascade Mountains for this adventure. They have a poster of the mountain and have been studying it, but they need more information about the terrain and elevations to decide on a hiking route.

Storyline: Students are introduced to the study of topography by building a model of the mountain landform. Students use the foam model of Mount Shasta to create and analyze a topographic map to gather information to plan a hike.

Phenomenon 2—Impact to Mount St. Helens: Students were looking at maps and images of the same volcanic mountain at different times. They observed some major differences in the landforms.

Storyline: After determining that two topographic maps are the same mountain, students gather information about Mount St. Helens before and after its devastating eruption in 1980. Students describe how the loss of stability caused major changes to the landform.

What kind of map do the hikers need and how will it help them to plan a route to the mountain peak?

FOCUS QUESTION:

How can we represent the different elevations of a mountain?

How did the mountain change and what caused the change?

Focus question and driving question are the same.

INV. 4 Changes to Landforms

Phenomenon 3—Earth-shaking events: Students in a classroom in South-Central Alaska dive under their desks when the building and its contents start shaking. They wait until the vibrations stop and evacuate with their teacher to the outdoors. People in the northern part of the state didn't feel the shaking.

Storyline: Students use maps to analyze the location data for active volcanoes and large earthquakes around the world and observe that many of these events occur in a pattern on the edges of the Pacific Ocean. Students use the online USGS Latest Earthquakes Map displaying the Earth's tectonic-plate outlines to gather information about the cause of these effects.

Phenomenon 4—Rock layers, flat and tilted: A family visits a natural outdoor site and finds large rock-layer formations with unusual stripes of red, pink, and yellow. Some of the layers are flat and others are tilted. In another site they discover fossils.

Storyline: Students interpret rock layers and the presence of fossils to determine the history of a place. They integrate the processes of weathering, erosion, deposition, canyon-forming (downcutting), and uplifting, which result in the rock formations we observe on the surface of Earth today.

Where do earthquakes and volcanoes occur around the world?

What do rock layers and fossils tell us about the history of a place?

Focus questions and driving questions are the same.

Content and Disciplinary Core Ideas	Practices and Crosscutting Concepts	NGSS PEs
ESS2.B: Plate tectonics and large-scale system interactions **ESS3.B:** Natural hazards • A topographic map uses contour lines to show the shape and elevation of the land. • The change in elevation between two adjacent contour lines is always uniform. • The closer the contour lines, the steeper the slope and vice versa. • The surface of Earth is constantly changing; sometimes those changes take a long time to occur, and sometimes they happen rapidly. • Catastrophic events have the potential to change Earth's surface quickly. • Scientists monitor volcanoes as they erupt to learn more about the earth-changing events.	**Science and Engineering Practices** Asking questions Developing and using models Analyzing and interpreting data Using mathematics and computational thinking Constructing explanations Engaging in argument from evidence Obtaining, evaluating, and communicating information **Crosscutting Concepts** Scale, proportion, and quantity Stability and change	**4-ESS2-2:** Analyze and interpret data from maps to describe patterns of Earth's features. **4-ESS3-2:** Generate and compare multiple solutions to reduce the impacts of natural Earth processes on humans.
ESS2.B: Plate tectonics and large-scale system interactions **ESS3.B:** Natural hazards **ESS1.C:** History of planet Earth • Most earthquakes and volcanoes occur in patterns near the edge of continents. • A world map with latitude and longitude lines is a tool to arrange tectonic data. The Ring of Fire has a majority of active volcanoes and earthquake activity. • Catastrophic events have the potential to change Earth's surface quickly. • Fossils in sedimentary rock provide evidence of organisms that lived long ago as well as clues to changes in the landscape and past environments. • Sedimentary rocks form in horizontal layers and those layers provide information about the geological history of an area. • The surface of Earth is constantly changing; sometimes those changes take a long time to occur and sometimes they happen rapidly. • Landforms are constantly changing due to weathering, erosion, deposition, downcutting, and uplifting.	**Science and Engineering Practices** Asking questions Developing and using models Analyzing and interpreting data Constructing explanations Obtaining, evaluating, and communicating information **Crosscutting Concepts** Patterns Stability and change	**4-ESS1-1:** Identify evidence from patterns in rock formations and fossils in rock layers to support an explanation for changes in a landscape over time. **4-ESS2-2:** Analyze and interpret data from maps to describe patterns of Earth's features.

Conceptual Flow of Module

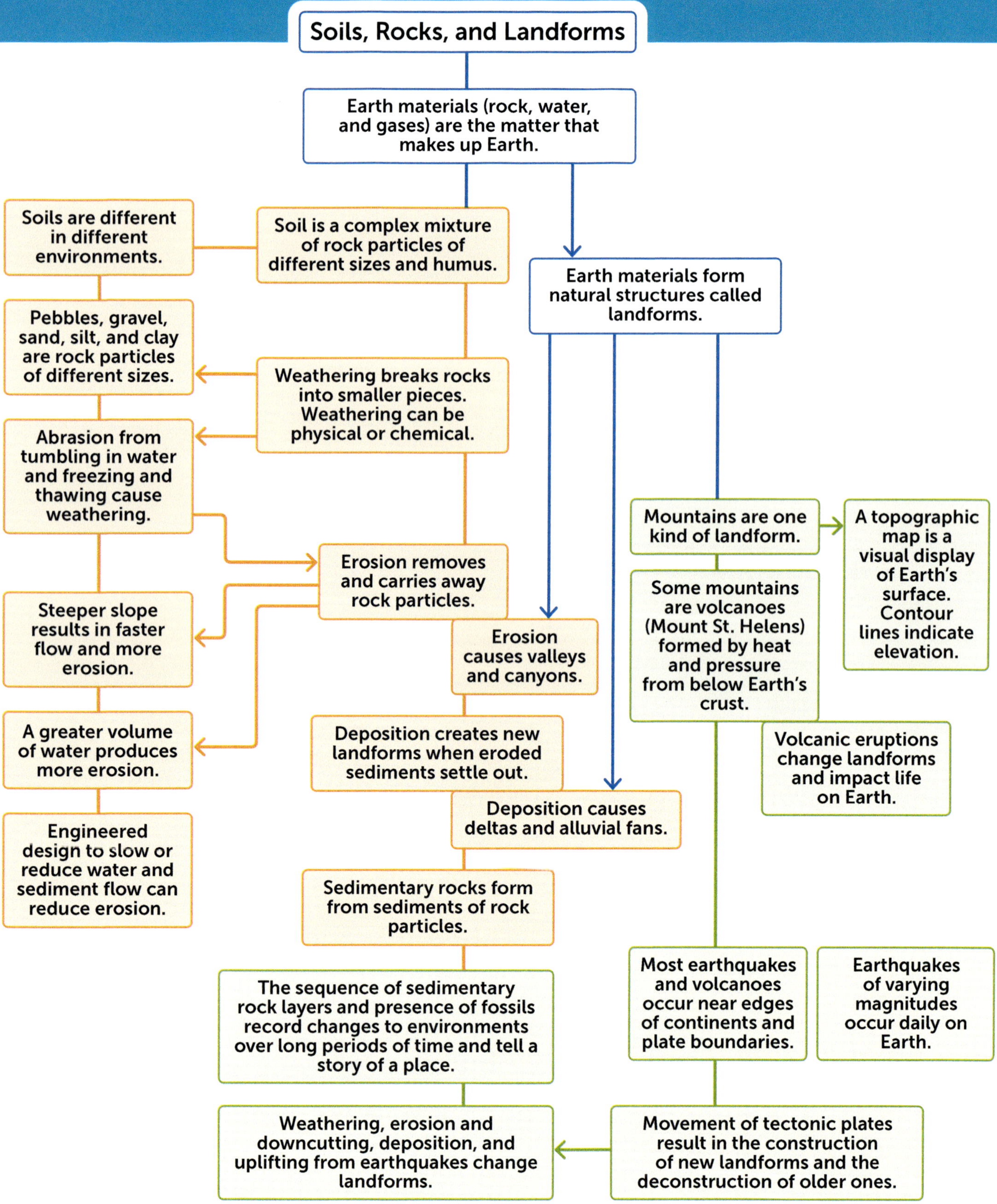

Driving questions emerge from four anchor phenomena. These questions support the learning progression by building on student ideas and core ideas over multiple investigations. We can see the coherence of ideas through the conceptual flow.

ANCHOR PHENOMENON 1 Investigations 1 and 2

Focus on weathering, erosion, and deposition—In Investigation 1, students analyze four soil samples and investigate the components to reveal differences that provide evidence as to the origin of each sample. They simulate weathering of rocks to model the formation of pebbles, gravel, sand, silt, and clay—the rock particles in soil. After exploring the origin of soil components, they investigate, in Investigation 2, how rock particles are transported from one place to another by water (erosion) and settle out through deposition to form new landforms or, on a smaller scale, a pile of earth material. Students apply this understanding to compare erosion barriers to reduce the impact of landslides on human structures.

Phenomenon 1: Pile of earth material on the schoolyard—What caused this pile of earth material to form?

ANCHOR PHENOMENON 2 Investigation 3

Focus on features of volcanic mountains—Investigation 3 begins with students finding out how to determine the elevation features of a mountain in order to plan a hike to the peak. Then they use their understanding of topographic maps to measure the change caused by an eruption to a volcanic mountain. Students are introduced to the Cascade Mountains as a chain of volcanic mountains. In Investigation 4, this chain will be linked to others when students map the location of active volcanoes.

Phenomenon 2: Impact to Mount St. Helens—How did the mountain change and what caused the change?

ANCHOR PHENOMENA 3 AND 4 Investigation 4

Focus on changes to landforms—Students find out where active volcano and earthquake activity are on Earth by mapping the locations of significant events to look for patterns. Students ask questions about tectonic events and use online resources to collect data to answer those questions. The final investigation provides virtual experiences with sedimentary rock formations in layers as students look for evidence to tell the geological history of a place.

Phenomenon 3: Earth-shaking events—Where do earthquakes and volcanoes occur around the world?

Phenomenon 4: Rock layers, flat and tilted—What do rock layers and fossils tell us about the history of a place?

OVERVIEW

Soils, Rocks, and Landforms

FOSS PATHWAYS TEACHING SCHEDULE

In the first year of implementation, follow the **critical pathway (shown in blue)**. In later years, plan how to expand learning opportunities to deepen and contextualize students' learning experiences. **Side Trips (shown in yellow)** provide ways to customize the curriculum to focus on environmental literacy and mathematics and scaling for grade 4. There are 29 sessions plus 5 Side Trip sessions.

Inv/Part	SOILS, ROCKS, AND LANDFORMS PATHWAYS	Number of Sessions
	Inv 1 Soils and Weathering	
Inv 1.1	*Survey* (optional)	
Inv 1.1	Soil Composition	3
Inv 1.2	Physical Weathering (1 optional session)	4
Side Trip 1	*Schoolyard Soils—Focus on environmental literacy*	2
Side Trip 2	*Chemical Weathering—Focus on carrying out investigations*	2
	Inv 2 Landforms	
Inv 2.1	Erosion and Deposition	3
Inv 2.2	Stream-Table Investigations	4
Inv 2.3	Reducing Erosion Impact	3
Inv 2.3	*I-Check 1–2*	1
	Inv 3 Mapping Earth's Surface	
Inv 3.1	Using Models and Topographic Maps	3
Side Trip 3	*Drawing a Profile—Focus on mathematics and scaling*	1
Inv 3.2	Mount St. Helens Case Study	1
	Inv 4 Changes to Landforms	
Inv 4.1	Mapping Earthquakes and Volcanoes	2
Inv 4.2	Interpreting Rock Layers	4
Inv 4.2	*Posttest*	1

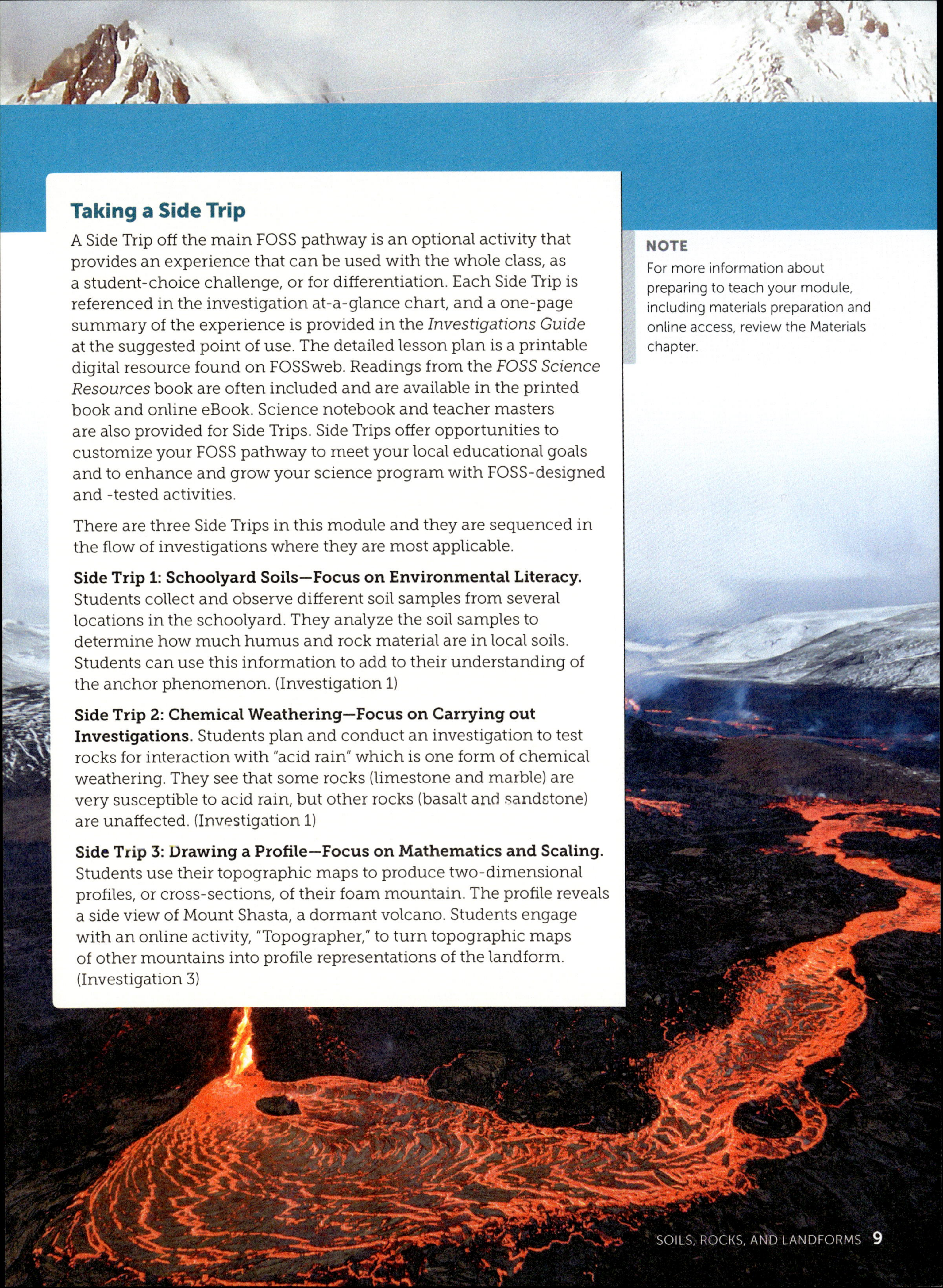

Taking a Side Trip

A Side Trip off the main FOSS pathway is an optional activity that provides an experience that can be used with the whole class, as a student-choice challenge, or for differentiation. Each Side Trip is referenced in the investigation at-a-glance chart, and a one-page summary of the experience is provided in the *Investigations Guide* at the suggested point of use. The detailed lesson plan is a printable digital resource found on FOSSweb. Readings from the *FOSS Science Resources* book are often included and are available in the printed book and online eBook. Science notebook and teacher masters are also provided for Side Trips. Side Trips offer opportunities to customize your FOSS pathway to meet your local educational goals and to enhance and grow your science program with FOSS-designed and -tested activities.

There are three Side Trips in this module and they are sequenced in the flow of investigations where they are most applicable.

Side Trip 1: Schoolyard Soils—Focus on Environmental Literacy. Students collect and observe different soil samples from several locations in the schoolyard. They analyze the soil samples to determine how much humus and rock material are in local soils. Students can use this information to add to their understanding of the anchor phenomenon. (Investigation 1)

Side Trip 2: Chemical Weathering—Focus on Carrying out Investigations. Students plan and conduct an investigation to test rocks for interaction with "acid rain" which is one form of chemical weathering. They see that some rocks (limestone and marble) are very susceptible to acid rain, but other rocks (basalt and sandstone) are unaffected. (Investigation 1)

Side Trip 3: Drawing a Profile—Focus on Mathematics and Scaling. Students use their topographic maps to produce two-dimensional profiles, or cross-sections, of their foam mountain. The profile reveals a side view of Mount Shasta, a dormant volcano. Students engage with an online activity, "Topographer," to turn topographic maps of other mountains into profile representations of the landform. (Investigation 3)

NOTE

For more information about preparing to teach your module, including materials preparation and online access, review the Materials chapter.

Soils, Rocks, and Landforms

FOSS INVESTIGATION ORGANIZATION

FOSS modules are divided into **investigations**. Investigations are divided into **parts**, and parts are broken down into sessions. Each investigation is associated with an **anchor phenomenon** (an observable event) and a related **driving question** that students investigate that provides general guidance and connects to student questions. Depending on the grade level and prior experiences, an anchor phenomenon might be introduced at the beginning of an investigation or after working with a few in-class phenomena.

Each part of each investigation is driven by a specific **focus question**. The focus question engages the students with the anchor phenomenon and signals the challenge to be met, mystery to be solved, or principle to be uncovered. The focus question, along with questions students ask, guide students' actions and thinking. Each part concludes with students recording a response to the focus question in their notebooks and connecting it to the anchor phenomenon.

The three-dimensional learning objectives for each investigation appear on the first page. They appear again on or near the last page of the investigation for the review.

Each investigation is summarized in the at-a-glance chart at the beginning of the investigation.

The investigation-specific **background** section for the teacher is presented in each investigation's introduction to present the anchor phenomenon and the focus questions. The learning objectives for that part are the first bulleted items in this section.

The **Materials** and **Getting Ready** sections provide scheduling information, an overview of the time required, and details about how to prepare the materials and resources for conducting the investigation (materials provided in the module, materials supplied by the school, and those available on FOSSweb).

The **Guiding the Investigation** is divided into 45-minute sessions, with sections with special features in the sidebars. These include several icons to remind you when a particular pedagogical method is suggested, as well as concise vocabulary lists, helpful hints, and ideas for supporting multilingual learners.

Sidebars also include callouts for **Science and Engineering Practices** and the **Crosscutting Concepts**. When there is an embedded assessment (notebook entry, response sheet, or performance assessment) the three-dimensional learning objective(s), including **Disciplinary Core Ideas**, involved in that assessment will be referenced in the sidebar.

Science and Engineering Practices

- Developing and using models

Disciplinary Core Ideas

- ESS2.A: Earth materials and systems

Crosscutting Concepts

- Cause and effect

Notes and **ELA Connections** appear in boxes in the sidebars. These notes serve one of two purposes. The first is the message you deliver to students. The second is designed to help you understand the science content and pedagogical rationale. Violet ELA Connection boxes identify English Language Arts objectives used in the lesson.

The **safety** icon alerts you to potential safety issues related to chemicals, allergic reactions, and the use of safety goggles.

The **vocabulary** icon indicates where students should review recently introduced vocabulary.

The **assessment** icon appears when there is an opportunity to assess student progress by using embedded or benchmark assessments. Some are performance assessments—teacher observations of science and engineering practices.

The **sense-making discussion** icon indicates when to engage in a student-centered discussion to help students develop their understanding of the science concepts. The chapter on sense-making discussions provides details on how to engage students with that concept in the context of the investigation.

The yellow **English Language Support note** provides a specific strategy to use to assist multilingual learners in developing science concepts.

The **Interpreting Information** section invites students to obtain, evaluate, and communicate information from the *FOSS Science Resources* articles and other media. Engaging in text and online activities help students deepen their understanding of science concepts.

The **Reflect on Phenomenon**, the last section in each part, provides opportunities for revisiting the focus question and how it relates to the anchor phenomenon. This section at the end of the investigation provides a review of all parts.

The last page of each investigation is **Differentiation Opportunities**. Listed on this page are student choice activities for group or individual projects. These opportunities are student-facing resources on FOSSweb. The last resource is the specific Home/School Connection that is used for learning at home.

ELA CONNECTION

This suggested strategy addresses this English Language Arts objective.

Apply word analysis skills in decoding words.

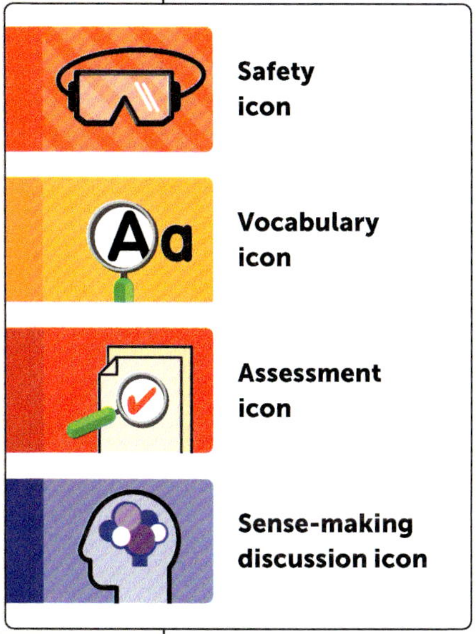

ENGLISH LANGUAGE SUPPORT

Have students help you make a class model as they explain their ideas. Be sure to label the parts with the new vocabulary words.

Soils, Rocks, and Landforms

THE ELEMENTS OF THE FOSS INSTRUCTIONAL DESIGN

FOSS is designed around active investigation that provides engagement with science concepts, science and engineering practices, and crosscutting concepts. Driving those firsthand investigations are instructional procedures that facilitate student sensemaking leading to understanding of phenomena and scientific habits of mind.

All FOSS investigations share **instructional design** strategies to provide consistent coherent engagement with science concepts in an iterative manner. There is variability in the implementation of the instructional design depending on the context and content of the science. Here is the general flow implemented to maximize every student's opportunity to learn.

- **Active investigation** in collaborative groups: firsthand experiences with phenomena in the natural and designed worlds and sense-making experiences in small groups and as a class to figure out phenomena. The design involves an anchor phenomenon with a driving question to activate prior knowledge and elicit student questions.
  - **context setting:** sharing prior knowledge, questioning, and planning investigations **(Engagement)**
  - **investigating:** doing, observing, collecting, and organizing data to figure out phenomena **(Exploration)**
  - **sensemaking:** analyzing data, constructing explanations, and communicating understanding about phenomena **(Explanation)**
- **Obtaining and interpreting information:** using media and local outdoor and community experiences to apply and extend the firsthand investigation to real-world contexts **(Elaboration)**
- **Reflecting using formative assessment:** monitoring progress continuously to inform student learning, and develop a growth mindset **(Evaluation)**

Engagement with real-world phenomena is at the heart of FOSS. Investigations present an anchor phenomenon and a **driving question**. In every part of every investigation, the investigative phenomenon is referenced implicitly in the focus question that guides instruction of each part and frames the intellectual work. Students record the focus question in their science notebooks, and after exploring and making sense of the phenomenon thoroughly, explain their thinking in words and drawings. At the end of each part and the investigation, students integrate their experiences through discussion of the driving questions for the entire investigation.

Reflecting using formative assessment
Figuring out phenomena using science notebooks
Applying science outdoors and in the local community
ACTIVE INVESTIGATION
Figuring out phenomena
Engaging in sensemaking and science-centered language development
Investigating ideas through technology
Obtaining and interpreting information through reading

Soils, Rocks, and Landforms

Instructional Design: Active Investigation

Figuring out Phenomena

Context setting: sharing prior knowledge, questioning, and planning. Active investigation requires focus. The context of an inquiry can be established with the introduction of a phenomenon and a driving question, a focus question, or a problem to be solved. The anchor phenomenon serves to activate prior knowledge, to elicit questions, and to connect the focus questions as students return to the driving question during the investigation.

Investigating: doing, observing, collecting, and organizing data. In the practice of science, scientists put things together and take things apart, observe systems and interactions, and conduct experiments. This is the core of science—active, firsthand experience with objects, organisms, materials, and systems in the natural world. In FOSS, students often conduct investigations in collaborative groups of four, with each student taking a role to contribute to the effort.

NOTE
Visit FOSSweb to download the **Science and Engineering Practices—Grade 4** chapter.

The active investigations in FOSS are cohesive and build on each other to lead students to a comprehensive understanding of phenomena. Through investigations and media, students gather meaningful data.

Data are the raw material from which scientific knowledge and meaning are synthesized. Students record and organize data so they will be easier to analyze. Tables allow efficient comparison. Organizing data in a sequence (time) or series (size) can reveal patterns. Students organize some data as graphs, providing a visual display of numerical data.

NOTE
Visit FOSSweb to download the **Sense-Making Discussions for Three-Dimensional Learning—Grade 4** chapter.

Sensemaking: analysis and constructing and communicating explanations. The most important part of an active investigation is extracting its meaning by figuring out phenomena. This constructive process involves prior knowledge, integrating new data, developing models, and discussing those ideas with others. Students share their explanations for phenomena, using evidence generated during the investigation to support their ideas. They conclude the active investigation by writing explanations in their science notebooks and proposing new questions.

Figuring out Phenomena Using Science Notebooks

The process of writing about their science experiences and communicating their thinking is a powerful learning device for students. Keeping a notebook helps students organize their observations and data, process their data, and maintain a record of their learning for future reference. The science notebook entries stand as credible and useful expressions of learning and form one of the core exhibitions of the assessment system. Look to the **Science Notebooks in Grades 3–5** chapter on FOSSweb for more details on how to use notebooks.

The notebook masters are reduced in size for placement into a bound composition book. Student work is entered partly in spaces provided on the notebook sheets and partly on adjacent blank sheets in the composition book.

NOTE
Visit FOSSweb to download the **Science Notebooks in Grades 3–5** chapter.

Engaging in Sensemaking and Science-Centered Language Development in Grades 3–5

The FOSS active investigations, science notebooks, FOSS Science Resources articles, and formative assessments provide rich contexts in which students develop and exercise communication. These elements are essential for effective instruction in both science and language arts. Students experience the natural world in real and authentic ways and use language to inquire, process information, and communicate their thinking about scientific phenomena. FOSS refers to this development of language process and skills within the context of science as science-centered language development.

The **Science-Centered Language Development in Grades 3–5** chapter on FOSSweb offers resources and strategies for you to use. The chapter describes how literacy strategies are integrated purposefully into the FOSS investigations, gives suggestions for additional literacy strategies that both enhance students' learning in science and develop or exercise English-language literacy skills, and develops science vocabulary with scaffolding strategies for supporting all learners. Standards for English Language Arts are included in the sidebars for each Guiding the Investigation.

NOTE
Visit FOSSweb to download the **Science-Centered Language Development in Grades 3–5** chapter.

Soils, Rocks, and Landforms

Obtaining and Interpreting Information through Reading

The *FOSS Science Resources* books, available in print and interactive eBooks, offer expository articles and biographical sketches. FOSS suggests that some readings be completed in class to connect to English Language Arts objectives. Other readings can be integrated into other areas of the curriculum. When language-arts skills and methods are embedded in content material that relates to the authentic experience students have had during the FOSS active investigations, students are interested, and they derive more meaning from the material.

Recommended strategies to engage students in reading, writing, speaking, and listening around the articles in the *FOSS Science Resources* books are included in Guiding the Investigation. In addition, a library of resources is described in the **Science-Centered Language Development in Grades 3–5** chapter on FOSSweb.

Investigating Ideas through Technology

The simulations and online activities on FOSSweb are designed to support students' learning at specific times during instruction. Digital resources include streaming videos that can be viewed by the class or small groups. Resources also include virtual investigations and tutorials that you can use for students who missed class or need to spend more time with the concepts.

There are many ways for students to engage with the digital resources—as individuals, in small groups, as a whole class, or at home with family and friends.

Applying Science Outdoors and in the Local Community

FOSS considers the entire school campus to be the science classroom. Taking your students into the schoolyard has three objectives: (1) the outdoor activities continue and extend the learning that starts in the classroom; (2) venturing outside provides opportunities to discover applications and everyday examples of the phenomenon being investigated; and (3) it connects students with nature and develops environmental literacy. When students are able to transfer knowledge of scientific principles to natural systems, they experience a sense of accomplishment. The **Taking FOSS Outdoors** chapter provides specific guidance for working with students outdoors in science.

NOTE
Visit FOSSweb to download the **Taking FOSS Outdoors** chapter.

Reflecting Using Formative Assessment

Scientists make sense of how the natural world works. They develop, test, and revise evidence-based explanations. Engineers apply that understanding to develop and improve solutions to human-made problems. The purpose of science instruction is to help every student become an effective sense maker and problem solver by doing what scientists and engineers do—at an age appropriate level—with guidance from their teachers.

The goal of assessment in the FOSS Program is to provide students with information they can use to improve their sensemaking and teachers with evidence of students' learning that they can use to adjust their instruction to meet the needs of every child.

The FOSS assessment system is designed to evaluate students' learning throughout a module. Embedded assessments focus on specific content and practices taught in that lesson. Benchmark assessments include the *Survey*, I-Checks, and *Posttest*. The *Survey* is given before instruction to provide information about students' prior knowledge. I-Checks can provide summative information about students' achievements, but may also be used to gather diagnostic information at the end of each investigation. Students can take the benchmark assessments online. Most items are automatically coded, and the system provides you with narrative reports, including alignment to NGSS.

To learn more about the FOSS Assessment System, see the **Assessment** chapter in this guide.

NOTE
Visit FOSSweb to download the Assessment masters and **Coding Guides** chapter.

DIVERSITY, EQUITY, AND INCLUSION

FOSS for All Students

The FOSS Program has been designed to maximize the science learning opportunities for all students, including those who have traditionally not had access to or have not benefited from equitable science experiences—students with special needs, ethnically diverse learners, English learners, students living in poverty, girls, and advanced and gifted learners. FOSS is rooted in a 40-year tradition of multisensory science education and informed by recent research on Universal Design for Learning and culturally and linguistically responsive teaching. Procedures to support students with special needs and students who are learning English are incorporated into the materials and strategies used with all students during the initial instruction phase. In addition, the **Access and Equity** chapter on FOSSweb provides strategies and suggestions for enhancing the science and engineering experiences for each of the specific groups noted above.

NOTE
Visit FOSSweb to download the **Access and Equity** chapter.

Throughout the FOSS investigations, students experience multiple ways of interacting with phenomena and expressing their understanding through a variety of modalities. Each student has multiple opportunities to demonstrate his or her strengths and thought process.

Your challenge is to provide appropriate follow-up experiences or enhancements appropriate for each student. For some students, this might mean more time with the active investigations or online activities. For other students, it might mean more experience and/or scaffolds for developing models, building explanations, or engaging in argumentation from evidence.

For some students, it might mean making vocabulary and language structures more explicit through concrete experiences or listening to the text read aloud to help them identify and understand relationships and connections.

For other students, it might be participating in individual projects or small-group investigations. It might be more opportunities for experiencing science outside the classroom in outdoor environments or identifying problems in the community and designing solutions.

ESTABLISHING A CLASSROOM CULTURE

Working in Collaborative Groups

Collaboration is important in science. Groups of researchers often contribute to the collection of data, the analysis of findings, and the preparation of the results for publication.

Collaboration is expected in the science classroom too. Some tasks call for everyone to have the same experience, either taking turns or doing the same things simultaneously. For other tasks, group members may have different experiences that they later bring together.

Research has shown that students learn better and are more successful when they collaborate. Working together promotes student interest, participation, learning, language development, and self-confidence. FOSS investigations use collaborative groups extensively.

For most activities in the elementary grades, collaborative groups of four in which students take turns assuming specific responsibilities work best. Groups can be formed randomly, or you can assemble groups to ensure diversity and inclusion. Groups can be maintained for extended periods of time or be reconfigured frequently.

Functional roles within groups can be determined by the members themselves, or they can be assigned in one of several ways. Each member in a collaborative group can be assigned a number or a color. Then you need only announce which color or number will perform a certain task. You can vary the colors and numbers in each investigation so that everyone has a chance to be a Getter, Starter, Recorder, or Reporter.

Norms for Sense-Making Discussions

Setting up norms for discussion and holding yourself and your students accountable creates a culture of productive talk in the classroom that supports engagement in the science and engineering practices. Students must feel free to express their ideas and to provide and receive criticism from others as they work toward understanding the disciplinary core ideas of science and engineering. Strategies for supporting academic discourse can be found in the **Sense-Making Discussions for Three-Dimensional Learning** and **Science-Centered Language Development** chapters on FOSSweb.

Functional roles used in FOSS:

- The **Getter** is responsible for handling the group's equipment.
- The **Starter** supervises setting up the equipment and makes sure that everyone gets a turn.
- The **Recorder** records the group's data and ideas, while ensuring everyone records the data in their science notebook.
- The **Reporter** reports the group's data to the class or writes it on the board or class chart and shares the main points during the class discussion.

This poster is an example of student responsibilities that the class discussed and adopted as their norms and is available on FOSSweb.

Soils, Rocks, and Landforms

NOTE

See the research of Carol Dweck and her book *Mindset: The New Psychology of Success.*

Collaborative Teaching and Learning

Collaborative learning requires a collective and individual growth mindset. A growth mindset is when people believe that their most basic abilities can be developed through dedication and hard work. As students work together to make sense of phenomena and develop their inquiry and discourse skills, you need to recognize and value their efforts to try new approaches and their willingness to make their thinking visible. Remind students that everyone in the classroom, including you the teacher, will be learning new ideas and ways to think about the world. Here are a few ways to help students develop a growth mindset for science and engineering.

- **Praise effort, not right answers.** When students are successful at a task, provide positive feedback about their level of engagement and effort (e.g., the efforts they put into careful observations, how well they organized and interpreted their data, the relevancy of their questions, how well they connected or applied new concepts, and their use of precise vocabulary). Provide feedback that encourages students to continue to improve their learning and exploring, such as: *Is there another way to approach this question? Have you thought about _____ ? What evidence is there to support _____ ?*
- **Foster and validate divergent thinking.** During sense-making discussions, frequently emphasize how important it is to share emerging ideas and to be open to the ideas of others in order to build understanding. Use think-alouds to model for students how you refine and revise your thinking based on new information. Make it clear that the point is not for them to show they have the right answer, but rather to help each other arrive at new understandings. Point out positive examples of students expressing and revising their ideas.

A classroom culture supports three-dimensional teaching and learning centers on collaboration. Collaborative groupings, materials management, and norms are structures you can put in place to foster collaboration. These structures, along with the expectations that students will be negotiating meaning together as a community of learners, creates a learning environment where students are compelled to work, think, and communicate like scientists and engineers.

Soils, Rocks, and Landforms

13,500
13,000
12,500
11,000

Soils, Rocks, and Landforms

Introduction

This chapter introduces the materials used in the FOSS Pathways investigations—where to find each type of material whether it be physical or digital, and what to do with the materials to prepare for teaching. Getting started and teacher preparation videos on FOSSweb complement this chapter.

Each part of each investigation has a list of materials to use for that part. The list identifies the materials needed for each student or group of students, for the class, or for the teacher. There are three main categories of materials on the list:

- Provided equipment
- School-supplied items
- FOSSweb resources

FOSS equipment kits are comprehensive and include the majority of equipment you need to teach the module. Except for a few common classroom items, FOSS kits include permanent equipment for one class of 32 students, working in collaborative groups of 4 and consumable equipment for three classes of 32 students.

FOSSweb teacher resources to support instruction include the e-Guide (for planning, teaching, and Side Trips) and videos and downloadable digital assets. These teacher-only digital assets are described as "Assessment" and "Module Resources."

FOSSweb has many other resources for both teachers and students including the module interactive eBook *FOSS Science Resources*, phenomenon and content videos and images, online multimedia activities, home/school connections, and family resources.

CONTENTS

NOTE

Check the Knowledge Base on ThinkLink to learn more about FOSSweb digital resources.

INVENTORY LIST

Provided Equipment

DRAWER 1 OF 3

Print Materials

Qty	Item	Equipment Condition
1	Teacher Print: *Soils, Rocks, and Landforms Investigations Guide, FOSS Science Resources: Soils, Rocks, and Landforms*[1]	
1	Poster, *Mount Shasta*	
1	Poster, *Mount St. Helens*	
1	Poster set, *Conservation*, 4/set	
2	Posters, *FOSS Science Safety* and *FOSS Outdoor Safety*	

NOTE

The teacher print is shipped separately. However, there is space in drawer 1 to store the *Investigations Guide* and teacher copy of the *FOSS Science Resources* book.

Items for Investigations 1

Qty	Item	Equipment Condition
1	Beaker, 100 mL	
8	Jars, plastic, with screw lids, 1/2 L	
100	Paper plates[2]	
40	Rock, conglomerate, pieces[2]	
40	Rock, granite, gray, pieces[2]	
4	Spoons, metal	
8	FOSS® trays, clear plastic	
25	Zip bags, 1 L	

DRAWER 2 OF 3

Items for Investigation 2

Qty	Item	Equipment Condition
1	Duct tape, roll[2]	
8	Meter tapes	
10	Rock, basalt, samples, small (fits in 12-dram vial)	
10	Rock, limestone, samples, small (fits in 12-dram vial)[2]	
10	Rock, marble, samples, small (fits in 12-dram vial)[2]	
10	Rock, sandstone, samples, small (fits in 12-dram vial)[2]	
8	Rulers, 30 cm	
8	Stream tables (tray, plastic)	
8	Water sources, flood, 1/2 L, with red dot	
8	Water sources, standard, 1/2 L, with blue dot	
16	Wood angles	

Items for Investigation 3

		Equipment Condition
8	Foam-mountain sets with dowel	
8	Maps, *Mount Shasta,* topographic maps	
8	Transparencies, *Mount Shasta* (hand-drawn from topographic map)	

Items for Investigation 4

10	Rock, banded sandstone	
10	Rock, shale	

DRAWER 3 OF 3

Shared Items

8	Basins	
9	Containers, 1/2 L	
8	Containers, with lids, 1 L	
50	Cups, plastic, 250 mL	
1	Pitcher	
8	Syringes, 50 mL	
40	Vials with caps, 12 dr.	

Consumable Items

1	Clay, powdered, brown, bags, 0.45 kg/bag (1 lb.)	
100	Cotton swabs	
1	Food coloring, set	
1	Gravel, bag, 2.3 kg/bag (5 lb.)	
2	Pebbles, bags, small, 0.5 kg/bag (1 lb.)	
1	Potting soil, 4 L bag (8 lb.) bag	
2	Sand, mountain, coarse containers, 1 kg/container (2 lb.)	
500	Self-stick notes	

Plastic box for earth material—permanent equipment

3	Clay, powdered, white, bags, 0.45 kg (1 lb.)/bag (stream table)[2]	
9	Sand, fine, bags, 1.35 kg (3 lb.)/bag (for stream table)[2]	

1 The student books, if included in your purchase, are shipped separately.

2 These items might occasionally need replacement.

INVENTORY LIST

School-Supplied Items

These are common classroom and school materials or consumable materials most easily provided locally for the investigation parts (not the Side Trips).

For all investigations

- Chart paper and marking pens
- Drawing utensils (pencils, crayons, marking pens)
- Glue sticks
- Science notebooks (composition books)
- Scissors
- Self-stick notes (for review sessions)

For outdoor investigations

1 Bag for carrying materials
32 Clipboards or cardboard pieces with binder clips

Investigation 1: Soils and Weathering

1 Glass bottle with tight-fitting cap
- Newspaper

1 Permanent black marking pen
- Safety goggles (optional)

1 Sturdy plastic bag
- Transparent tape
- Vinegar, white, about 1.2 L
- Water

8 Zip bags, large

Investigation 2: Landforms

1 Dustpan and broom
- Gloves and face masks for dust (optional)
- Newspaper
- Paper towels
- Water

Investigation 4: Changes to Landforms

- Blue and red colored pencils
- Rulers or straight edge (optional)
- Poster downloaded from USGS website, "This Dynamic Planet"

TO TEACH THE MODULE

Preparation

Be sure to review the *Preparing to Teach the Module* video on FOSSweb for helpful tips to help you set up your classroom for teaching FOSS.

Some preparation is required each time you use the kit. Doing these things before beginning the module will make daily setup quicker and easier.

1 Inventory consumable materials

A number of items in the kit are consumable. Some of these items will be used up during the investigations (food coloring, gravel, gray powdered clay, self-stick notes, potting soil), and others will wear out and need replacement (duct tape, paper plates, evaporation dishes, stream table earth material ingredients, and some of the rocks). Before throwing items out, consider ways to recycle them and get your students involved in this process.

2 Prepare or check four soil types (Investigation 1)

In Investigation 1, Part 1, students investigate four soil samples. Each group will need 1 vial about 2/3 full of each soil type.

Use the materials in the kit (gray powdered clay, gravel, pebbles, and potting soil for humus) to measure and mix these soil samples following the directions in Step E of Getting Ready. The recipe will make enough for one class.

There is enough materials in the kit to make the soil samples for three classes. Use only what you need for your class.

NOTE

The four soil types are:
River delta (1)
Mountain (2)
Desert (3)
Forest (4)

3 Separate gray granite and conglomerate

Separate the pieces of gray granite and conglomerate into two groups: rocks that students will tumble (16 rocks of each kind), and rocks that will not be tumbled but used as a control (4 rocks of each kind). This allows multiple classes to tumble the rocks and still have some unmarred rocks for comparison.

4 Check vials

The vials are reusable and should be rinsed and dried before going back into the kit. Check them for cracks and replace when necessary.

NOTE

The stream-table earth materials are shipped in a small plastic box, not in a FOSS drawer. The bags of white (not gray) clay and fine sand in this plastic box are considered permanent equipment for use in the stream table.

5 Use the FOSS tray

The clear plastic FOSS tray was designed for this module. Each group of four students uses one tray. It is designed to hold four vials in corner sockets and four evaporating dishes in the larger openings. It also serves as an organizer for rock samples. The trays nest for storage and stack for space efficiency while soil samples are settling during investigations. Vials, without caps, placed in the corner sockets serve as support columns.

6 Check earth material (Investigation 2)

The total volume of the earth material needed for one stream table is about 1 L. If the kit has been used before, check that each earth-material bag contains about 1 L of the sand-clay mixture.

If the bags seem to have the correct volume, add to each bag one 12-dram vial of white clay. White clay is usually lost during the stream-table investigations.

If the bags do not have enough earth material, pour all of them into a big bucket. Add a cup of white clay to the bucket and mix. Use a 1 L container to measure a volume of the mixture for each of the eight bags. If you still come up short, use the recipe below to make another bag. Additional bags of fine sand (1.35 kilograms [kg]/bag) and white clay (0.45 kg/bag) can be ordered from Delta Education.

a. Add one plastic cup of white powdered clay (about 100 grams [g]) to a 1.35 kg bag of fine sand.

b. Mix the materials in the zip bag until they are uniformly distributed.

7 Become familiar with types of clay

There are two kinds of clay in the kit used for different purposes.

- The gray powdered clay is used in Investigation 1, Part 1, to make the soil samples. This clay is consumable.
- The white powdered clay is used in Investigation 2, Part 1, as part of the stream table earth material mixture. This earth material can be used over and over again and is considered permanent equipment.

8 Provide magnifying lenses

If possible, provide several magnifying lenses for students to use during the module to observe the rocks.

9 Check condition of containers and water sources

A variety of containers are included in the kit. Check that all containers noted below are clean and in good condition. Be sure to use the correct container for each investigation by referring to the materials list for each part and viewing the teacher preparation video.

All the plastic containers in the module are intended to be rinsed and reused.

- Containers, 1/2 L
- Containers, 1 L
- Plastic cups
- Vials, 12 dram
- Water source, standard, with blue dot (with small hole in bottom)
- Water source, flood, with red dot (with larger hole in bottom)
- Vials, 12 dram

NOTE

The four soil types are:
River delta (1)
Mountain (2)
Desert (3)
Forest (4)

10 Check foam mountains (Investigation 3)

Each foam-mountain set consists of six foam pieces and one dowel. Make sure the sets are complete. If students have used water-based ink pens in tracing the shapes, the edges of the foam pieces may have ink on them. The foam can be washed with soap and water to remove the ink.

11 Check transparencies

The kit contains 8 transparencies of teacher master 5, *Foam-Mountain Topographic Map*. Make sure they are clean and ready for use. Use teacher master 5 to make replacements if needed.

12 Obtain safety equipment (optional)

Check your district regulations regarding the use of safety goggles. Students will be using small amounts of vinegar in Investigation 1. Safety goggles can be purchased from Delta Education.

In Investigation 2, students work with a sand and clay mixture in the stream table experiments. For students with respiratory issues, face masks and gloves can be used.

FOSSweb Resources

Teacher Only

Be sure to review the *Preparing to Teach a Module for Grades 3–5* video on FOSSweb for helpful tips to help you set up your classroom for teaching FOSS.

Videos and downloadable digital resources support instruction. These teacher-only resources are organized on FOSSweb in these categories.

NOTE
Students can take benchmark assessments online where most items are automatically coded for the teacher. See the Assessment chapter for more information about the FOSSmap online assessment for grades 3–8 and view the FOSSmap tutorials available on FOSSweb Knowledge Base.

ASSESSMENT

1 Assessment Masters

Benchmark assessments include the *Survey*, I-Checks, and *Posttest*. They can be administered in hard copy or online with FOSSmap. Find out more about these benchmark assessments by reading the Assessment chapter in this *Investigations Guide*.

2 Assessment Coding Guides

These are the answer sheets and coding guides for all benchmark assessments. The latest coding guides are identified by the school year on the first page.

3 Assessment Charts

The teacher notes, checklists, and records are available as PDFs and some are available as spreadsheets. These resources include:

- *Embedded Assessment Notes* for embedded assessments
- *Performance Assessment Checklist* for embedded assessments
- *Assessment Record* for benchmark assessments

NOTE
Check the Knowledge Base on ThinkLink to learn more about FOSSweb digital resources for students and teachers including eBook, videos, multimedia activities, home/school connections, student choice activities, and other family resources.

MODULE RESOURCES

1 Teaching Resources

The three teaching resources for materials are

- **Teacher Videos:** View the Soils, Rocks, and Landforms Module Teacher Preparation Video and the Anchor Phenomenon Overview Video for the module to supplement the *Investigations Guide* to plan for each lesson.
- **Teaching Slides:** These downloadable and editable slides can be used to facilitate each part of each investigation with the class. They are available as Google slides in English and Spanish.
- **Important Module Updates:** These are announcements of changes to materials or updated procedures.

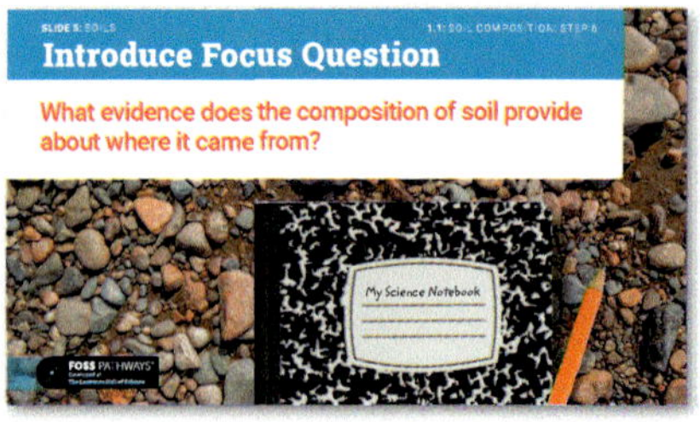

2 Notebook and Teacher Masters

You will need to print or make copies of science notebook sheets before each investigation. You can download, print, and project electronic copies of the sheets from FOSSweb. They are available as PDFs and Google sheets in English and Spanish.

- **Teacher masters,** full sheets
- **Notebook masters,** 2-half sheets/page
- **Notebook masters for display** (form-filled)
- **Notebook masters for Google Classroom**

3 Materials and Kit Information

The FOSS developers suggest using a central materials distribution system. You organize the materials for a session at a single location, the materials station. As the investigation progresses, a member of each group, the Getter, gets materials as they are needed, and another Getter returns the materials at the end of the session.

These digital material resources will make instruction more efficient.

- **Module Kit Contents** (same as pages 2–3 of this chapter)
- **Grade-Level Kit—Grade 4** (the inventory list for the 3-module equipment kit)
- **Equipment Photo Cards** (English and Spanish)

  Use a pocket chart to display the equipment photo cards as reference for students as they gather needed items from the materials station for each part.
- **Word Wall Cards** (English and Spanish)

  Word wall cards for each part can be downloaded and printed. As the module progresses, add new vocabulary words to a word wall or pocket chart, and model writing and responding to focus questions in a class notebook using those words.
- **Posters** (English and Spanish)

  All posters and card sets are provided in the kit as hard copy. In addition, digital formats are provided. Spanish posters are only available digitally. Review the posters with students and post them in the classroom.

  Two safety posters are included in the kit, one for science indoors and one for working outdoors. Use the four *Conservation* posters to discuss the importance of conserving natural resources.

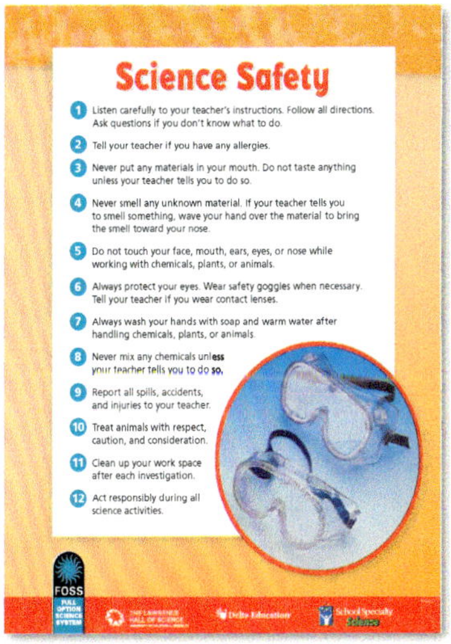

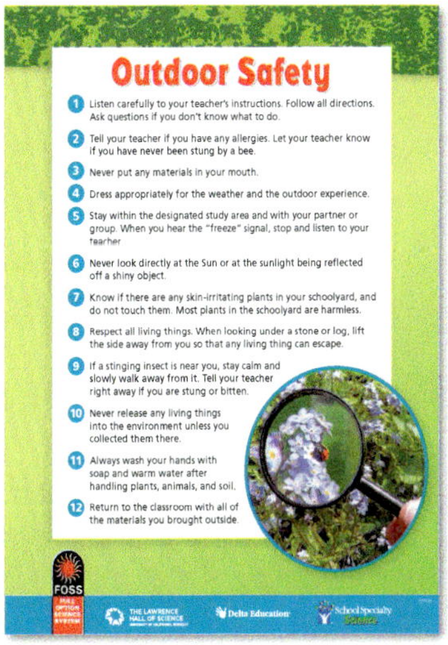

Care, Reuse, and Recycling

The items in the module have been selected for their ease of use and durability. When you finish teaching the module, inventory the equipment carefully. Note the items that were used up, lost, or broken, and immediately arrange to replace the items. Use a photocopy of the materials list (the Provided Equipment Inventory List), and put your marks in the "Equipment Condition" column.

Refill packages and replacement parts are available for FOSS by calling Delta/School Specialty at 1-800-258-1302 or by using the online replacement-part catalog.

Standard refill packages of consumable items are available from Delta/School Specialty. A refill package for a module includes sufficient quantities of all consumable materials (except those provided by the teacher) to use the kit with three classes of 32 students.

Here are a few tips on storing the equipment after use.

- Inventory and bag the small items.
- Wash vials, cups, and evaporation dishes. Be sure all equipment is dry before returning it to the kit.
- Rocks and minerals that have been in vinegar should be rinsed thoroughly with mild soapy water and dried.
- Sort rocks and minerals into labeled bags.
- Rinse and dry the wood angles before storing them.
- Dry earth material thoroughly and store in a plastic bag or a bucket.
- Secure the lids on all containers of materials and on the food coloring.
- Place each foam mountain and its dowel in its own bag.
- Check the posters to ensure that they are dry and flat in the kit.
- Check quantity of consumables, and order more if necessary.

Any items that cannot be reused for the particular FOSS investigation might be usable in another part of the curriculum. Before throwing items out, consider ways to recycle them, and get your students involved in this process.

- **Safety Data Sheets** (SDS)

  There are updated specifics for all items requiring safety information. If you have questions regarding any SDS, call Delta/School Specialty at 1-800-258-1302 (Monday–Friday, 8 a.m.–6 p.m. EST).

Soils and Weathering

INVESTIGATION 1

Soils and Weathering

Phenomenon Storyline

Pile of earth material on the schoolyard: On a walk around the schoolyard, students discover some changes at the edge of the playground. They observe a new pile of sand and "dirt" near the bottom of a small hill.

What caused this pile of earth material to form?

Students observe some changes at the edge of the playground, a new pile of earth material. To figure out how the pile formed, students gather data to determine the composition of soils from four different locations. They discover that not all soils are the same and explore the processes of weathering that break down rocks to form soils. They plan and conduct tests to determine the effects of different types of weathering on different materials, such as the rocks found near playgrounds. This first investigation focuses on the origin of the earth material (where did the material come from?) and Investigation 2 focuses on the movement of the material (what caused it to end up on the schoolyard in that pile?)

If possible, use an eroded section of your schoolyard or image from a similar spot in the community to make this scenario a local phenomenon.

THREE-DIMENSIONAL LEARNING OBJECTIVES

- Students observe, record, and analyze data to reveal patterns about the different materials found in soil. **(Part 1)**
- Students use patterns from firsthand observations and media to engage in argument about the source of different types of soil. **(Part 1)**
- Students plan and carry out investigations to determine the effects of weathering on different earth materials such as those found near schoolyards. **(Part 2)**

INVESTIGATION 1

Soils and Weathering
At a Glance

Investigation Part	Sessions	Practices and Crosscutting Concepts
Introduce anchor phenomenon 1—Pile of earth material on the schoolyard		
PART 1 Soil Composition **FOCUS QUESTION: What evidence does the composition of soil provide about where it came from?** Students observe and compare four different soils. They learn that soils are composed of essentially the same types of materials (inorganic earth materials and humus), but the amounts of the materials vary. Students speculate on the origin of each of four soils: mountain, desert, river delta, or forest. This firsthand introduction to soils provides a lens for students to use to consider the scenario of a new pile of earth material on the schoolyard at the bottom of a hill.	**SESSION 1** Steps 1–12 **SESSION 2** Steps 13–20 **SESSION 3** Steps 21–24 *A session is 45 minutes.	**Practices** • Asking questions • Planning and carrying out investigations • Analyzing and interpreting data • Constructing explanations • Engaging in argument from evidence • Obtaining, evaluating, and communicating information **Crosscutting Concepts** • Patterns • Systems and system models
PART 2 Physical Weathering **FOCUS QUESTION: What causes big rocks to break into smaller rocks?** Students begin to explore how large masses of rock break into smaller pieces. They tumble rocks and freeze water to see how these two types of physical weathering cause rocks to break into smaller rock pieces. This provides them with information about the small rock particles they found in different soil types and a way to make observations of new samples of earth materials.	**SESSION 1** Steps 1–9 **SESSION 2** Steps 10–14 **SESSION 3** Steps 15–18 (Step 18 optional) **OPTIONAL SESSION 4** Steps 19–20 **SESSION 5** Steps 21–23 **SIDE TRIP 1** **SIDE TRIP 2**	**Practices** • Developing and using models • Planning and carrying out investigations • Analyzing and interpreting data • Constructing explanations • Engaging in argument from evidence • Obtaining, evaluating, and communicating information **Crosscutting Concepts** • Cause and effect

Performance Expectation

4-ESS2-1. Make observations and/or measurements to provide evidence of the effects of weathering or the rate of erosion by water, ice, wind, or vegetation.

Disciplinary Core Ideas	Interpreting Information	Assessment
ESS2.A: Earth materials and systems **ESS2.E:** Biogeology • Soils can be described by their properties. • Soils are composed of different kinds and amounts of earth materials and humus. • Living things affect the characteristics of soil. • The composition of soil provides evidence of its origins.	**Science Resource Book** "Parts of Soil" **Other Media** Anchor Phenomenon Resource Video 1	**Embedded Assessment** *Survey* **Embedded Assessment** Science notebook entry **NGSS Performance Expectation addressed** 4-ESS2-1
ESS2.A: Earth materials and systems **ESS2.E:** Biogeology • Weathering is the breakdown of rocks and minerals at or near Earth's surface. • The physical-weathering processes of abrasion and freezing and thawing and the pressure of tree roots (living things) break rocks and minerals into smaller pieces.	**Science Resource Book** "Weathering" "Making Concrete" (optional) "Earth Materials in Art" (optional) **Other Media** Video: *Freezing Glass Bottle* (optional) Tutorial: "Weathering" (optional) Video: *Chemical Weathering* (optional Side Trip 2)	**Embedded Assessment** Response sheet **NGSS Performance Expectation addressed** 4-ESS2-1

INVESTIGATION 1

PART 1 Soil Composition

Science and Engineering Practices

- Asking questions
- Planning and carrying out investigations
- Analyzing and interpreting data
- Constructing explanations
- Engaging in argument from evidence
- Obtaining, evaluating, and communicating information

Disciplinary Core Ideas

- ESS2.A: Earth materials and systems
- ESS2.E: Biogeology

Crosscutting Concepts

- Patterns
- Systems and system models

clay	rock
earth material	sand
gravel	silt
humus	soil
pebble	system

Teacher Background

Phenomenon Storyline

LEARNING OBJECTIVES:

- **Students observe, record, and analyze data to reveal patterns about the different materials found in soil.**
- **Students use patterns from firsthand observations and media to engage in argument about the source of different types of soil.**

The anchor phenomenon for the first two investigations is based on the following scenario. On a walk around the schoolyard, students discover some changes at the edge of the playground. They observe a new pile of sand and "dirt" near the bottom of a small hill.

What caused this pile of earth material to form?

This part provides initial experiences for students to begin their conceptual development of weathering of solid earth materials. Here are some things students will find out to connect to the storyline.

- Soils are not all the same. Soils from different environments (mountain, desert, river delta, forest) have different amounts of various earth materials.
- The materials near the bottom of the hill look similar to those materials found in some soils.

FOCUS QUESTION

What Evidence Does the Composition of Soil Provide about Where It Came From?

Soil is the loose top layer of Earth's surface. It's made of small particles of rock, bits of plant material, and living organisms.

Earth's surface was once solid rock. Over a very long time, the rocky surface broke down into smaller and smaller pieces. People have assigned names to different-sized rock particles. Rocks between the size of grapes and golf balls are called pebbles. The next smaller class of rocks, between the size of rice grains and grapes, is called gravel. Bits of rock smaller than rice grains, down to salt grains, are called sand. Smaller than sand, but still observable with simple magnification and detectable by feel as fine grit, are particles called silt. The tiniest category of rock pieces, too small to see with a hand lens, is clay. When clay is moistened and rubbed between the fingers, it feels slippery. When a significant mass of rock particles of various sizes accumulates, it forms the mineral portion of soil. But to rise to the level of soil, the rock material needs to be infused with organic material (humus).

Five factors determine the ever-changing characteristics of soil: (1) the type of rock source, (2) the climate, which influences weathering, (3) the topography, (4) the actions of living things, and (5) the passage of time.

Organisms create and enrich soil. Rock material without organic matter (humus) is not really soil at all. When insects, worms, and vertebrates dig and burrow in the soil, they stir in leaves and other organic material, and their waste products and decomposing bodies are a critically important contribution to the fertility of soils. The development of soil requires time—time for the components to assemble, time for organisms to make the rock material fertile, and time to establish a healthy ecosystem of organisms to perpetuate the vitality of the soil.

Soil provides a medium of support for plants. Plants invade the soil with their roots. Their roots extract vital nutrients from the soil—water and minerals. The nutrient minerals come from simple compounds that are the products of decomposition of dead plants and animals as well as from chemicals extracted directly from the rock material.

Different soils have different characteristics based on the relative amounts of different-sized rock particles and amount of humus. Soil can be studied as a system composed of different parts that work together.

The conceptual flow for this module starts with looking for sources of the sand and dirt by examining soil. Students find that soil is a complex mixture of rock particles of several sizes (pebbles, gravel, sand, silt, clay) and decomposing organic matter (humus). Different environments on Earth tend to have generalizable soil types based on the percentage of the basic soil components. Students begin their investigation of soils from different environments in this part and complete it in Investigation 2, Part 2, Step 17. Students will find out these characteristics of different soils:

- River delta soils are typically high in clay and silt.
- Mountain soils tend to have large pieces of earth material, not a lot of humus.
- Desert soils tend to be mostly sand and gravel, not a lot of humus.
- Forest soils tend to have a fairly even distribution of mineral particle sizes—sand, silt, and clay—as well as high concentrations of organic humus.

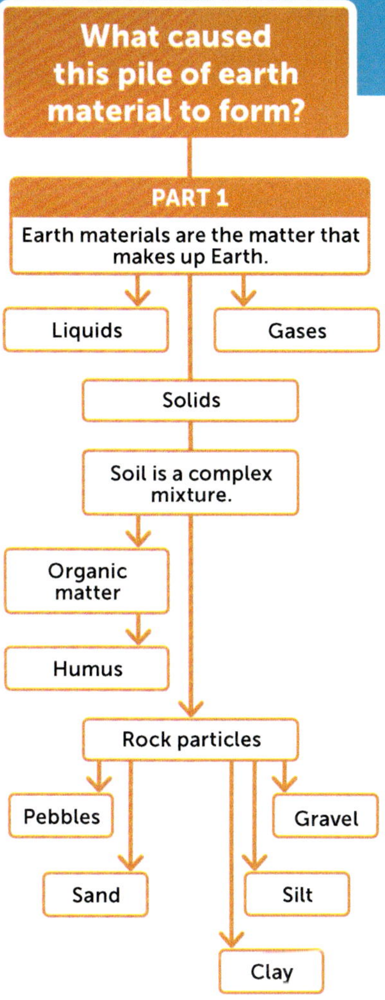

PART 1 Soil Composition

Materials

PROVIDED EQUIPMENT

For each student

- **1** *FOSS Science Resources: Soils, Rocks, and Landforms* "Parts of Soil"

For each group

- **1** Basin
- **4** Vials with caps, 12 dram
- **1** FOSS tray
- **1** Syringe
- **4** Paper plates
- **5** Self-stick notes
- **1** Container, 1/2 L

For the class

- **4** Self-stick notes
- **4** Metal spoons
- **4** Zip bags, 1 L
- **1** Beaker, 100 mL
- **4** Containers, with lids, 1 L
- • Pebbles
- • Gravel
- • Sand, coarse (or "mountain sand")
- • Clay, gray, powdered
- • Potting soil (humus)
- **1** Pitcher
- • *Science Safety* and *Outdoor Safety* posters
- • *Conservation* posters

SCHOOL-SUPPLIED ITEMS

For the class

- Science notebooks (composition books)
- Transparent tape
- Water

FOSSWEB RESOURCES

For each student

- *Survey* (or have students take the *Survey* online)
- Notebook sheet 1, *Soil Observations*
- Notebook sheet 2, *Soils in Vials*
- Teacher master 1, *Letter to Family*

For the class

- Word wall cards and equipment cards
- Teaching slides
- Anchor Phenomenon Resource Video 1

For the teacher

- *Embedded Assessment Notes*
- *Assessment Record* (for *Survey*)

Soil Observations

Soil 1	Soil 2

Soil 3	Soil 4

Soils in Vi

1

3

Location

Location

Location

Location

Location

FOSS PATHWAYS
© The Regents of the University of California
Can be duplicated for classroom or workshop use

SOILS, ROCKS, AND LANDFORMS MODULE
Investigation 1: Soils and Weathering
No. 1—Notebook Master

No. 2—Notebook Master

No. 1–2—Notebook Masters

Letter to Family

Science News

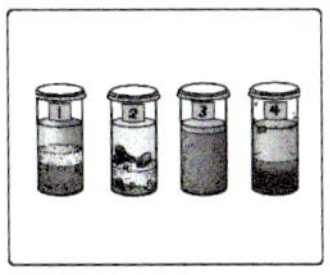

Dear Family,

Our class is beginning a new science unit using the **FOSS Soils, Rocks, and Landforms Module**. We will observe soils and conduct investigations about how the weathering of rocks provides the basic ingredients that form soil. Using stream-table system models we will investigate how the processes of erosion (transportation) and deposition (settling out) move earth materials to shape new landforms.

We will become familiar with many types of landforms—plateau, mountain, plain, valley, canyon, alluvial fan, delta, and more. We will study volcanic mountains using physical models, topographic maps, and photographs to obtain information to help us understand the shape of the land they represent.

We will figure out how changes in Earth can be due to slow processes, such as erosion, as well as rapid processes, such as landslides, volcanic eruptions, and earthquakes. And we will map the locations of a selection of active volcanoes and the epicenters of major earthquakes, analyze and interpret the data, and look for patterns. We will go on virtual field trips to observe rock layers of different sedimentary rocks and learn how to read clues that tell us the history of a place.

You can increase your child's understanding and interest in earth science by asking them to talk about the investigations we do at school. Take a walk around your neighborhood and look for the effects of weathering, erosion, and deposition. Talk about the impacts of natural Earth processes on your community, state, or region (earthquakes, floods, volcanic eruptions) and what is being done to reduce the impacts. Engage with a virtual Erosion Engineering challenge on FOSSweb and compare methods to reduce the impact of landslides on human structures.

Watch for Home/School Connection sheets that I will be sending home from time to time. These describe ways your whole family can look more closely at soils, rocks, and landforms around your home.

We're looking forward to weeks of fun with earth science! If you have questions or comments or have expertise you would like to share with the class, please drop me a note. You can get more information on this module by going to www.FOSSweb.com.

FOSS PATHWAYS
© The Regents of the University of California
Can be duplicated for classroom or workshop use

SOILS, ROCKS, AND LANDFORMS MODULE
Investigation 1: Soils and Weathering
No. 1—Teacher Master

No. 1—Teacher Master

Embedded Assessment Notes

Investigation ____, Part ____ Date ________

Got it!

Concept

Concept

Reflections/Next Steps

Embedded Assessment Notes

Investigation ____, Part ____ Date ________

Got it!

Concept

Concept

Reflections/Next Steps

FOSS PATHWAYS
© The Regents of the University of California
Can be duplicated for classroom or workshop use

SOILS, ROCKS, AND LANDFORMS MODULE
Embedded Assessment Notes
No. 1—Assessment Master

No. 1—Assessment Master

Assessment Record: Survey/Posttest Date ________

Student names	1	2	3a	3b	4a	4b	5	6	7a	7b	8	9	10	11a	11b

FOSS PATHWAYS
© The Regents of the University of California
Can be duplicated for classroom or workshop use

SOILS, ROCKS, AND LANDFORMS MODULE
Assessment Record
No. 4—Assessment Master

No. 4— Assessment Master

PART 1 Soil Composition

NOTE

The Teacher Preparation video and the Anchor Phenomenon Overview video for this module will give you a big picture of the instruction and preparation for *Soils, Rocks, and Landforms*.

Getting Ready

Schedule

Survey

3 Sessions

Session 1: Steps 1–12
Session 2: Steps 13–20
Session 3: Steps 21–24

Review

- Background for Part 1
- FOSSweb Resources for this part
- Reading: "Parts of Soil"
- Teacher Preparation Video
- Anchor Phenomenon Overview Video
- Anchor Phenomenon Resource Video 1

Print or Copy

For each student:

- *Survey*, or schedule it online through FOSSmap
- Notebook sheet 1, *Soil Observations*
- Notebook sheet 2, *Soils in Vials*
- Teachers master 1, *Letter to Family*, or send it home online

For the teacher:

- *Embedded Assessment Notes*
- *Assessment Record*

Prepare Material

- Use of science materials — A
- Science notebooks and word wall — B
- Working with multilingual learners — C
- Provide water — D
- Mix soils — E

Plan for Assessment

- Review Coding Guide for *Survey* (identical to *Posttest*) — F
- Review Step 19, "What to Look For" in the notebook entry — F

PREPARATION DETAILS

A Use of science materials

If this is your first use of this module, become familiar with the equipment. Review the provided equipment (kit) inventory list. Review the Materials chapter for recommendations about organizing and distributing materials, planning for school-supplied materials, and setting up the classroom environment for science.

NOTE

Review the Preparing to Teach a Module Grades 3–5 video on FOSSweb that supports the Materials chapter. The short video will help you set up a classroom environment that promotes science teaching and learning.

B Science notebooks and word wall

Plan for student notebooks by reviewing the Science Notebooks in Grades 3–5 chapter. A class notebook using chart paper will serve as a reference for students. Print the word wall cards from FOSSweb and plan for a class word wall.

C Working with multilingual learners

Suggestions for English Language Support appear in the sidebar throughout Guiding the Investigation to provide additional strategies for making science accessible for students who are learning English. Refer to the Science-Centered Language Development chapter.

D Provide water

Half fill a 1/2 L container with water for each group. Keep the water containers and syringes out of the reach of students until Step 10.

E Mix soils

Use the materials in the kit and the 100 mL beaker to make four different soil types following the table. Mix the ingredients thoroughly in a basin before transferring them to the 1 L containers. Label them 1 through 4, but **do not** include the names of the simulated environments. (Use the gray clay, not white.)

Soil	Clay (mL)	Sand (mL)	Gravel (mL)	Pebbles (mL)	Humus (mL)	Simulated environment
1	150	110	110	0	30	river delta
2	30	30	150	150	30	mountain
3	30	150	180	0	30	desert
4	15	30	110	15	225	forest

Place the numbered containers of soil and a metal spoon for each at the materials station. This recipe makes enough "soil" samples for one class use. Each group needs 2/3 of a vial of each soil.

F Plan for assessment

The *Survey* is part of the FOSS benchmark assessments. It will help you determine what students already know and what they need to learn.

The science notebook entry is part of the embedded, formative assessment system. Review the "What to Look For" ahead of teaching the lesson.

INVESTIGATION 1

PART 1 Soil Composition

Guiding the Investigation

SESSION 1 **45 MINUTES**

1 Introduce the anchor phenomenon

Introduce the anchor phenomenon by describing and sharing the scenario or by going on a walk in the schoolyard to make firsthand observations.

On a walk around the schoolyard, students discover some changes at the edge of the playground. They observe a new pile of sand and "dirt" near the bottom of a small hill.

What caused this pile of earth material to form?

Have students share their thinking and prior knowledge with a partner. Ask students what questions they have about the pile of earth material on the schoolyard and what ideas they might try to better understand how it formed. When students suggest that the earth materials is soil, continue to the next step.

NOTE
Use the Anchor Phenomenon Resource Video 1 or an alternative local resource for this introduction. An alternative might be to take a brief walk in the schoolyard. Ask students what they observe in the video or what they observed on a walk in the schoolyard.

2 Introduce *earth materials*

Tell students,

Today we're going to begin a study of ***earth materials*** *and how the shape of Earth's surface changes over time. Earth materials are the various solids, liquids, and gases that make up planet Earth. The earth materials we're going to look at today are in the* ***soils*** *in these containers.*

Ask a student to describe to the class what is in one of the containers of soil. Again, students may call it "dirt," but use the word soil to talk about what is in each container (students will develop a specific definition of soil over the course of this activity). Ask students to name places where soils are found.

ENGLISH LANGUAGE SUPPORT
Begin a word wall titled "Soil Composition" and write the words "earth materials" and "soils." Explain that "composition" in this context means what something is made of.

3 Go on a soil-observation walk (optional)

Take your students to several locations in the schoolyard or neighborhood to discover that soils are everywhere and may look different from one place to another. Make this trip outside brief. This will give students an opportunity to look at soils from a fresh viewpoint and start asking questions.

Tell students that they are going to go on a quick walk. Their task is to look for soil (and for sources of a pile of earth materials) and consider the question *Where is soil and what is it made of?* Explain rules for expected behavior and set boundaries, referring to the *FOSS Outdoor Safety* poster. Have students get coats as needed and head outdoors.

NOTE
If you used the video in Step 1 to introduce the scenario, now is a good time to make the activity relevant to students in their schoolyard or neighborhood. You may want to provide metal spoons and zip bags to get soils samples.

Encourage students to share their observations and questions as you tour the area. If needed, ask,

- ➜ *Where should we look for soil?*
- ➜ *What kinds of materials do you see in the soil? Is anything growing?*
- ➜ *Are there soil areas where nothing is growing?*
- ➜ *How is this soil the same as the soil we saw over there? How is it different?*
- ➜ *What questions do you have that you might want to investigate?*

When you see soil near lawn, pavement, or buildings, ask,

- ➜ *Where did this soil on top of the lawn or pavement come from?*
- ➜ *Do you think there is soil underneath the lawn, the pavement, or the building?*

4 Return to class

Have students return to the classroom. If they collected soil samples, set those aside for now. Ask students to discuss what they observed.

- ➜ *Where did we find soil?*
- ➜ *Is soil the same in every location? Do you see any patterns?*
- ➜ *Why might geologists (scientists who study Earth) and other people be interested in studying soil?*

5 Introduce science notebooks

Distribute science notebooks to students. Have students write the numbers 1–10 on the outside corner of the first ten pages, using the front and back of each page. Reserve pages 1–3 for a table of contents. Refer to the Science Notebooks in Grades 3–5 chapter on FOSSweb.

6 Introduce the focus question

FOCUS QUESTION

What evidence does the composition of soil provide about where it came from?

Project or write the focus question on the board while you read it aloud.

- ➜ *What evidence does the composition of soil provide about where it came from?*

Have students write the date and focus question in their notebooks. Distribute copies of notebook sheet 1, *Soil Observations*. Students should glue the sheet on page 4 of their notebook. Or, if you prefer not to use the notebook sheet, students can divide a blank page into four sections, one section for each soil.

Science and Engineering Practices

- Asking questions

Crosscutting Concepts

- Patterns

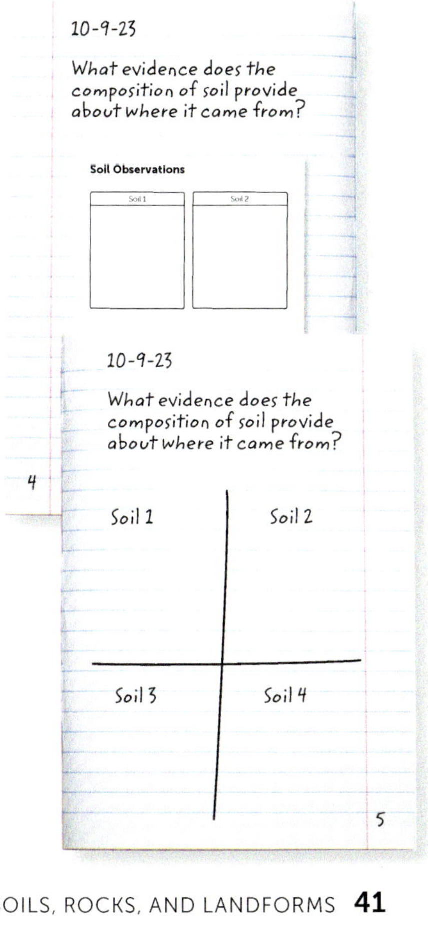

PART 1 Soil Composition

SESSION 1

MATERIALS FOR STEP 7

- Containers of soils, 1–4
- Paper plates
- Empty vials and caps
- Self-stick notes
- Transparent tape
- Scissors

7 Collect soil samples

Have Getters go to the materials station and pick up four paper plates, four empty vials and caps, and five self-stick notes. Each person in a group labels one paper plate and its matching vial. Attach a self-stick note with a number ("1," "2," "3," or "4") to the plates. Use one quarter of a numbered self-stick note and a small piece of transparent tape to put labels on vials. If necessary, model the procedures for numbering the plate and vial.

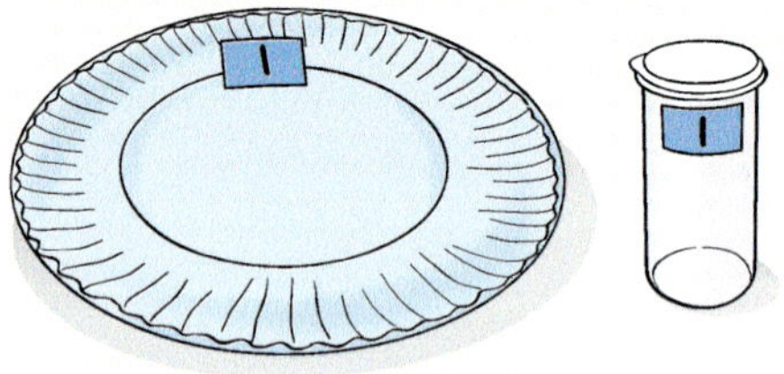

In tag-team fashion, one person at a time from each group goes to the materials station and fills his or her numbered vial two-thirds full of soil. Be sure students match the number on the vial to the number on the soil container and bring it back to the table. If different groups start with different numbers, it will alleviate the crowd at the materials station.

Science and Engineering Practices

- Planning and carrying out investigations

8 Observe "dry" soils

Have students start the investigation by making observations. Have them pour a soil sample onto a paper plate, matching the number on the vial with the number on the paper plate. Students should observe and record properties such as color, shape of materials, particle size, and textures. Students should draw what the samples look like in their notebooks.

As students complete their observations of one soil, have them carefully exchange paper plates until all group members have the opportunity to observe and record observations for all four soils.

Don't discard the paper plates. You should be able to reuse one set of paper plates for the entire module.

NOTE

The purpose here is for students to notice that soils are made of the same basic materials: different sizes of rock or mineral and humus.

9 Discuss materials that make up soils

Call for attention or gather students together for a discussion with their notebooks. Ask,

➜ *What different kinds of materials did you find in the soils?*

List the materials that students name on the board.

When students attempt to name the different sizes of **rock** material, help them by giving each size a name. Describe the size pattern.

- **Pebbles:** 6–64 mm
- **Gravel:** 2–6 mm
- **Sand:** 1–2 mm
- **Clay** or **silt:** rock particles the size of powder and dust with particles less than 1 mm

When students make note of organic material in the mix, introduce the word **humus**. Explain that plants and animals living in or near the soil eventually die and decay. They become part of the soil. Humus is the dead material that becomes part of soil.

Take a poll. Have students consider the soil that is in front of them. Name each material on the list and ask students to raise their hands if they found that material in the soil they are looking at now. Then ask,

➜ *How are the four soils the same?* [They are all made of small rock particles and humus (dead plant and animal materials).]

Explain that this is a pattern and helps to identify and describe different types of soil.

➜ *How are they different?* [The soils have different amounts of each kind and size of material.]

Establish the pattern that all soils are made of humus (dead plant and animal material) and different sizes of rock material. But the soils may have different amounts of each kind and size of material.

10 Add water to separate components

Have students fold their paper plates to create a pathway to transfer the soil samples back into the numbered vials. Tell students that water can be a useful tool to sort the different materials that make up soils. Have Getters go to the materials station and pick up a syringe and 1/2 L container of water for the group.

If this is students' first time using a syringe, show them how to use it. Remind students not to aim syringes at other students. Have students use syringes to transfer 25 mL of water from the container into each soil-filled vial. A quick and easy way to measure 25 mL of water is by pulling the plunger out until the notch is level with the top of the barrel.

As you go around the room and watch students add the water, point out the bubbles rising from the soil as the water soaks in.

Ask,

➜ *What do the bubbles indicate?* [As the water seeps in, it pushes air out. The bubbles are evidence that there are air spaces between the particles of soil.]

ENGLISH LANGUAGE SUPPORT

Write the rock-particle-size categories on the word wall as a reference. Tape a zip bag of each material next to its name.

Crosscutting Concepts

- Patterns

MATERIALS FOR STEP 10

- Syringes
- Containers of water

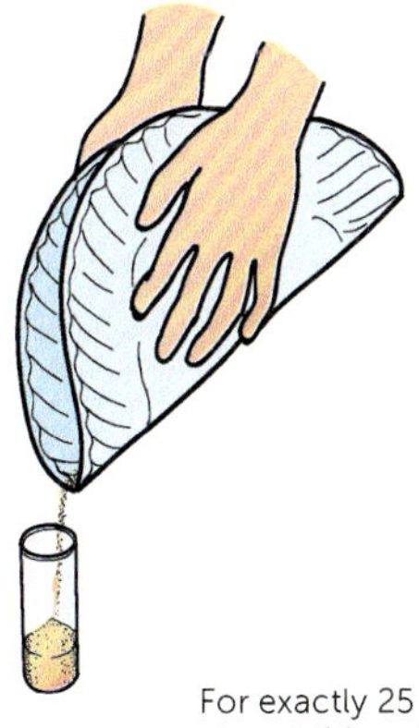

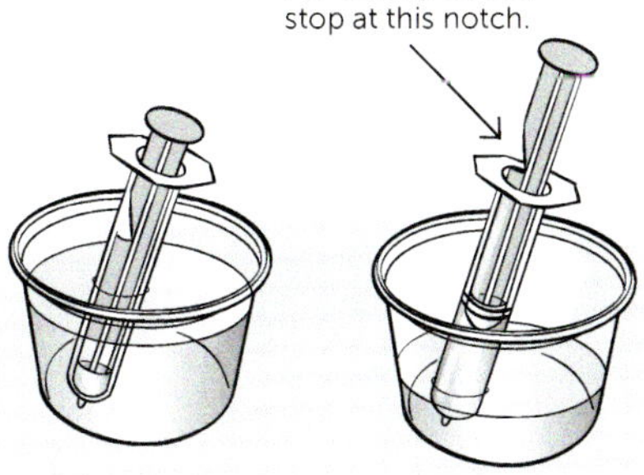

SESSION 1

When everyone has added water, have students cap the vials tightly and shake the contents, holding the cap in place as they shake.

After a minute or two of shaking, ask students to place the vials on their desks and observe them undisturbed for 3–5 minutes. Students should look closely as the layers of soil materials form. They should not touch the vials so they can note the order in which the materials settle in the vial.

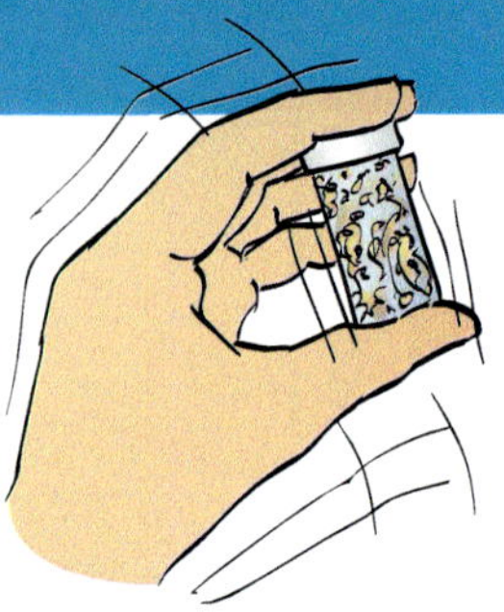

11 Analyze and interpret observations

Review the earth materials and humus that make up soil. Ask,

→ *Which materials settled out first after shaking the vials?* [Larger materials settle first. The smaller materials, especially silt or clay, can stay suspended in the water for quite a long time.]

→ *Where is the humus?* [Students may see pieces of humus floating at the top of the water as well as mixed in with the earth materials, giving it a black color.]

→ *We observed bubbles when you added water to the soil in the vial. What does that tell you about soil?* [There must be air spaces between the soil particles. When water is added, it pushes the air out from the spaces, and we see the air rise as bubbles.]

Science and Engineering Practices

- Analyzing and interpreting data

NOTE

Pebbles in the mountain soil may look like they are sitting on top of finer earth material, but they do settle first. The pebbles fall to the bottom immediately, and then the finer materials fill in the gaps between the pebbles, so it looks like the layering is different.

12 Settle soil vials overnight

Have students put all four vials in the corners of a FOSS tray, label the tray with their group's number or names, and set the trays in a location where they won't be disturbed overnight. Have students return the rest of the materials to the materials station.

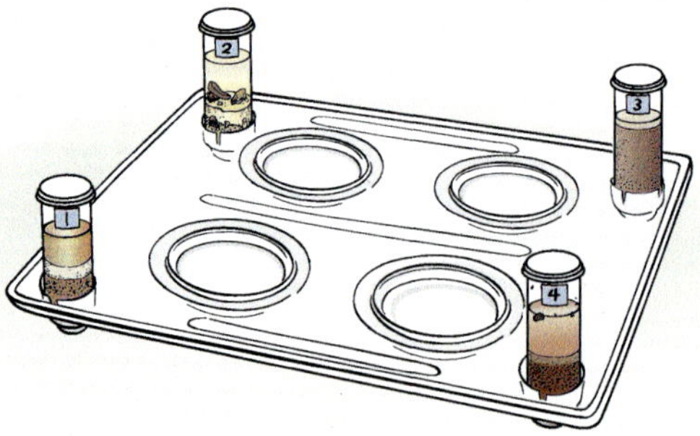

You will need a place to store the soil samples while they sit overnight. The FOSS trays are designed to stack on vials without lids. If you have room for each group to place their tray on a counter or table, then have the students keep the lids on the vials of soil and water.

ENGLISH LANGUAGE SUPPORT

Encourage students to talk about their observations with their partners and/or record them in their notebooks.

SESSION 2 | 45 MINUTES

13 Observe the settled vials

Have one Getter from each group pick up four copies of notebook sheet 2, *Soils in Vials*, at the materials station. Have another Getter pick up the group's tray with its vials, moving carefully to the table so as not to disturb the settling. Ask students to record what they see in the vials by making drawings on notebook sheet 2, *Soils in Vials*.

14 Make sense of collected data

Have students analyze and compare the four different soils their group has been observing. Ask,

- → *Do all the vials contain the same amount of soil?* [All started with the same amount.]
- → *Do all the vials have the same number of layers?* [They should.]
- → *Let's look for patterns in the series of vials. Are all the layers of similar earth materials the same size (depth) in all the different soils? For example, do all the soils have the same size sand layer? Do they have the same amount of humus or organic matter? What patterns, if any, do you observe?*

Use this discussion to help students define soils. Soils are a mixture of the same kinds of basic rock materials and humus, but the amounts and sizes of the basic materials vary.

Science and Engineering Practices
- Analyzing and interpreting data

Crosscutting Concepts
- Patterns

15 Connect to the anchor phenomena

Refer to the pile of earth materials phenomenon and ask students to apply what they have observed so far about soil to answer their questions. Use their questions to guide the discussion and help students construct explanations.

- → *How were you able to separate the earth materials in the vial?* [Added water, shook, and let settle.]
- → *What ideas do you have about where the earth material came from?*
- → *What did you observe that makes you think that?*
- → *How might this happen in nature?*

Science and Engineering Practices
- Constructing explanations

16 Discuss where the soils came from

Tell students that the soil samples they have been observing represent the kinds of soils that might be found in four locations. The samples represent the desert, mountain top, river delta, and forest soils. Write these locations on the board or chart paper.

Have students discuss in their groups about why soils might be different from one location to another. Then have students make a preliminary claim about where each vial of soil came from. Students can write their preliminary claims below each vial on notebook sheet 2.

ENGLISH LANGUAGE SUPPORT
Provide pictures of the different location and add them to the word wall. Small pictures are found in the teaching slides.

INVESTIGATION 1

PART 1 Soil Composition

SESSION 2

Have pairs of students present to each other their preliminary ideas (claims) and provide an explanation for their thinking. Tell students they are going to carry out more investigations to learn more about how soils form and other processes that shape Earth in the next few weeks. After doing more investigations, they will then have an opportunity to refine their ideas and engage in argument from evidence about their claims.

clay	rock
earth material	sand
gravel	silt
humus	soil
pebble	

17 Review vocabulary

Review key vocabulary introduced in this part. If those words are not on the word wall, add them now.

18 Respond to the focus question

Have students respond to the focus question in their notebooks.

➜ *What evidence does the composition of soil provide about where it came from?*

For students who need scaffolding, provide a sentence frame such as: *Soil is composed of ______. In our investigation, we found out ______.*

19 Assess progress: notebook entry

Have students hand in their notebooks open to the page on which they answered the focus question. You might also want to take a quick look at students' preliminary claims to help you prepare for the next set of lessons.

Science and Engineering Practices

- Analyzing and interpreting data

Disciplinary Core Ideas

- ESS2.A: Earth materials and systems
- ESS2.E: Biogeology

Crosscutting Concepts

- Patterns

WHAT TO LOOK FOR

- *Based on analysis and interpretation of firsthand observations, students write that soil is composed of different sizes of rock (e.g., sand, gravel, pebbles) and humus (decaying plants and animals).*
- *The components of the soil form layers (patterns) when each sample is mixed with water and allowed to settle. The amount of each component provides some evidence for where the soil came from. The soil with the biggest rocks might be from the mountains. The soil with the most humus might be from the forest.*

20 Save one set of soil samples

Save one group's set of vials as a reference when they refine their claims in Investigation 2, Part 2, Step 17.

Have students return all equipment to the materials station. Clean out the rest of the vials. Set up two basins, one empty and the other filled with water. Students can dump the soil and water into the first basin and rinse the vials in the second basin. All of the soil and water should be recycled appropriately outdoors.

Interpreting Information

SESSION 3 **45 MINUTES**

21 Read "Parts of Soil"

Give students a few minutes to look at and discuss the cover of the *FOSS Science Resources* book. Then have them examine and discuss how the table of contents and the glossary can help them locate information. Students will read articles throughout the module in order to obtain, evaluate, and communicate information to reinforce or extend the classroom firsthand experiences to the real world.

Tell students that "Parts of Soil" is an expository article that focuses on the various components that make up soil. First, have students preview the text by looking at and discussing the photographs and the diagram with a reading partner. Have students discuss what information they can obtain from these illustrations. [Soil comes from the ground; it is composed of different-sized rocks that can be categorized by their size.]

Refer to the title and suggest that as students read they look for new information that provides details on the parts of soil. They should also be thinking about how their observations in class compare to what they read in the article. Read the article aloud or have students read independently. For reading strategies to support diverse learners, see the Access and Equity and the Science-Centered Language Development chapters.

ELA CONNECTION

These suggested strategies address these English Language Arts objectives.

Determine the main idea of a text; summarize the text.

Interpret information presented visually.

Science and Engineering Practices

- Obtaining, evaluating, and communicating information

PART 1 Soil Composition

SESSION 3

Review the words in bold and then ask students to describe soil in terms of a **system**. Write the word *system* on the word wall.

➜ *What are the component parts of soil as a system?* [Rocks, humus, live organisms, water, and air.]

➜ *How do the soil components work together as a system?* [Living organisms live and die in the soil; they mix up the soil allowing space for air and water to move; humus provides nutrients and helps retain water.]

To support comprehension, make (or have students make) a word web for the components of soil. Have students include the word web in their notebook.

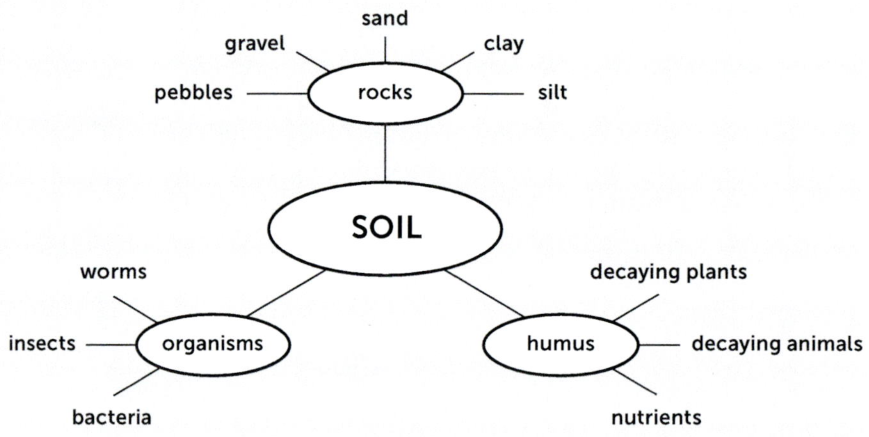

Crosscutting Concepts

- Systems and system models

ELA CONNECTION

These suggested strategies address these English Language Arts objectives.

Determine the meaning of general academic domain-specific words or phrases.

Compare and contrast a firsthand and secondhand account of the same topic.

Engage in collaborative discussions.

Parts of Soil

Have you ever dug a hole in the ground? What did you remove from the hole? **Soil**. Sometimes people call it dirt. But the scientific name for the layer of diggable material that covers planet Earth is soil.

If you pick up a handful of soil and look at it closely, you might be able to see and feel what soil is made of. Soil is mostly made of several sizes of **rock**. You might see pebbles and smaller pieces of gravel. Soil usually contains sand. **Particles** of sand are teeny tiny rocks. Some pieces of rock are even smaller than grains of sand. These smaller pieces are called **silt**. The smallest pieces of rock are clay particles. **Clay** particles are too small to see, but you can feel them. Clay feels slippery when it is wet.

Digging in soil

2

This means that soil is a mixture of different-sized rocks (pebbles, gravel, particles of sand, and even smaller particles of silt and clay) as well as water and air. Rocks, air, and water are all **earth materials**.

Soil is mostly made of different-sized rocks, water, and a

Rock Size Cha

Particle name	Average
pebble	
gravel	
sand	
silt	
clay	invisible to bare eye

3

Make Connections Ask students to compare their soils in the photographs to the soils they have been investigating in class. How are they the same and different?

Soil also contains **organic material**. Organic material is the remains of dead plants and animals. Plant roots grow in the soil, and animals dig into the soil. When plants and animals die, their remains break down into tiny pieces. These pieces become part of the soil called **humus**. Humus provides **nutrients** for plants. Humus also helps the soil **retain** water.

Worms are good for soil and help plants grow. Worms burrow through the soil. As they move, worms mix the soil and make passageways for air and water. Worm waste also adds nutrients that are good for soil and plants.

Worms break down organic material in the soil as they eat.

4

Not all soils are alike. Some kinds of soil have more humus. Some soils have more clay. Some have more sand, pebbles, and gravel.

Questions for Reflection

1. What differences do you see in the soils shown above?
2. Where do you think these soils are found?

5

Apply Understanding Have students share their observations and ideas about how worms affect soil. Ask guiding questions such as, "Why do you think worms are found in humus-enriched soil? Do the actions of worms benefit humans? In what way?"

ELA CONNECTION

These suggested strategies address these English Language Arts objectives.

Explain how an author uses reasons and evidence to support particular points in a text.

Engage in collaborative discussions.

Report on a text in an organized manner, using appropriate facts and relevant, descriptive details to support main ideas or themes; speak clearly at an understandable pace.

Science and Engineering Practices

- Constructing explanations
- Engaging in argument from evidence

NOTE

Go to FOSSweb and look for the Science and Engineering Practices—Grade 4 chapter for details on how to engage students with the practice of engaging in argument from evidence.

SESSION 3

22 Discuss the reading

To check for understanding, have students discuss these questions in their groups and then ask a Reporter from each group to share the group's responses.

➜ *What did you already know about soil? What was new information?*

➜ *What is humus?* [Decayed dead plants and animals.]

➜ *What does humus do for soil?* [Provides nutrients and helps soil retain water.]

Refer students to the statement on page 4, "Worms are good for soil and help plants grow." Ask students to explain how the author uses reasons and evidence to support that statement. Encourage them to cite details and examples from the text.

Ask students to look closely at the soils on page 5 and to describe different kinds of soil. [Some soil has more clay, some more sand, some more pebbles and gravel, some more humus.]

Discuss the questions at the end of the article to give students more experience constructing explanations with evidence.

➜ *What differences do you see in the soils shown above?* [Color of soil, length of grass roots, the one on the left seems to have more sand and the one on the right has more humus.]

➜ *Where do you think these soils are found?* [Accept all answers but ask for reasoning.]

Call on Reporters to share their group's answers and allow other group Reporters to engage in argument by either agreeing or disagreeing or offering alternative ideas. To support collaborative discussions, provide sentence frames such as:

- *I think ______ because ______.*
- *I agree with ______ because ______.*
- *I disagree with ______ because ______.*
- *I would like to add that ______.*
- *Why do you think ______?*
- *What is your evidence that ______?*

Reflecting on a Phenomenon

23 Construct a partial explanation of the anchor phenomenon

This a good time to have students record their thinking about the pile of earth material on the schoolyard. Have them create a section in their notebook where they can begin to construct their explanation of the phenomenon and can return to it in Investigation 2 to refine their explanation. Show the visual in the Anchor Phenomenon Resource Video 1 as needed to facilitate students' thinking. Review the driving question:

→ *What caused this pile of earth material to form?*

Encourage students to apply what they've learned about soils so far to figuring out what might be in the earth materials in the anchor phenomenon.

24 Share notebook entries

Conclude Part 1 or start Part 2 by having students share their notebook entries. Ask students to pair up with a partner to:

- discuss the different kinds of materials that were found in the four different soils;
- share their ideas about which soil came from which location based on the pattern of the layers—desert, river delta, mountain, forest. Students should support their ideas with reasoning;
- share what new questions they have for further investigations.

ENGLISH LANGUAGE SUPPORT

Give students a moment to discuss their ideas with a partner before writing in their notebooks. If needed, provide a writing frame such as:

In order to figure out what might be in the earth materials, first we _____. Then, we _____. So far, we know that _____. I think _____. I'm still wondering _____.

Crosscutting Concepts

- Patterns

Science and Engineering Practices

- Developing and using models
- Planning and carrying out investigations
- Analyzing and interpreting information
- Constructing explanations
- Engaging in argument from evidence
- Obtaining, evaluating, and communicating information

Disciplinary Core Ideas

- ESS2.A: Earth materials and systems
- ESS2.E: Biogeology

Crosscutting Concepts

- Cause and effect

Aa

abrasion
conglomerate
expand
freeze
granite
model
physical weathering
weathering

Teacher Background

Phenomenon Storyline

CONTINUED

LEARNING OBJECTIVE:

- **Students plan and carry out investigations to determine the effects of weathering on different earth materials such as those found near schoolyards.**

This part encourages students to extend their ideas about the formation of soil and the production of rock particles that can become new landforms. To figure out how the pile of earth materials on the schoolyard formed, students plan and conduct tests to determine the effects of types of weathering on different materials such as the rocks found near schoolyards.

Here are some things students will find out to connect to the storyline.

- Rocks break into smaller pieces. Those pieces are given names based on their size. Some of those pieces become the parts of soil.
- The physical-weathering processes of abrasion and freezing break rocks and minerals into smaller pieces.

FOCUS QUESTION

What Causes Big Rocks to Break into Smaller Rocks?

Early in its history, Earth's surface was one continuous expanse of molten rock. In time, it cooled to form a continuous, solid rocky crust. There was no soil, and there were no small bits of rock—no sand, gravel, or silt. As Earth continued to cool, the rock contracted and began to crack. Rock pieces began to break away and tumble downhill. When rocks tumble, they collide with other rocks. The force of impact breaks large rocks into smaller pieces. When rocks break up as a result of physical forces, the process is called physical weathering. Rocks don't necessarily require collision to break apart; even friction or abrasion causes them to break into increasingly smaller pieces.

When Earth cooled enough for liquid water to condense, water started to flow over Earth's surface. Moving water pushes rocks around and causes more frequent and powerful impacts. If water invades cracks in rocks and then freezes, the expanding water causes a process called frost wedging. Rocks high in the mountains continually weather due to ice expansion.

Large pieces of rock dislodged from cliffs fall, and the force of the fall can break a large slab of rock into many pieces. Ice and gravity all play significant roles in the physical weathering of rock. We can develop models to study this natural process.

Rocks are composed of minerals. Minerals are chemical compounds. The kinds and concentrations of minerals in a source rock significantly influence the soils they produce. The ways the source rock weathers (breaks down) determines the size of the particles in the soil. Topography determines how the weathered particles are distributed—how they get laid down and mixed. Steep gradients encourage smaller particles to travel in flowing water. Flat basins allow silt and clay particles to accumulate.

In Part 2, students grapple with the question of where the small rocks in soil come from. Students are introduced to the idea that small rocks are the broken pieces of large mountain-sized rocks. Those large masses of rock are reduced to boulders by physical weathering caused by frost wedging (freezing and thawing) and further reduced to sand, silt, and clay by abrasion when rocks fall and tumble downhill in landslides and moving water.

In Side Trip 1, students can extend and apply what they have learned in the classroom to soil samples from the schoolyard.

Students read about another form of weathering that reduces rock to tiny particles—chemical weathering. During chemical weathering, water or acids dissolve the minerals in a rock. The process of reacting removes portions of rock, one molecule at a time. The removal of minerals can weaken a large mass of rock so much that it crumbles into small pieces. The products of the chemical reaction can dissolve, creating a solution that eventually precipitates or crystallizes in a new location. Chemical weathering, particularly by acid rain, is responsible for a significant amount of the breakdown of rock.

Some rocks and the minerals that make up those rocks are affected by acid rain more than other rocks. Rocks with the mineral calcite, such as limestone and marble, react more in the presence of acid rain than rocks such as granite, basalt, conglomerate, and sandstone.

When plants, particularly trees, send their roots into cracks in rocks, pressure from the growing roots can break apart the surrounding rocks. This is primarily physical weathering, but for some kinds of plants it can involve chemical weathering as well.

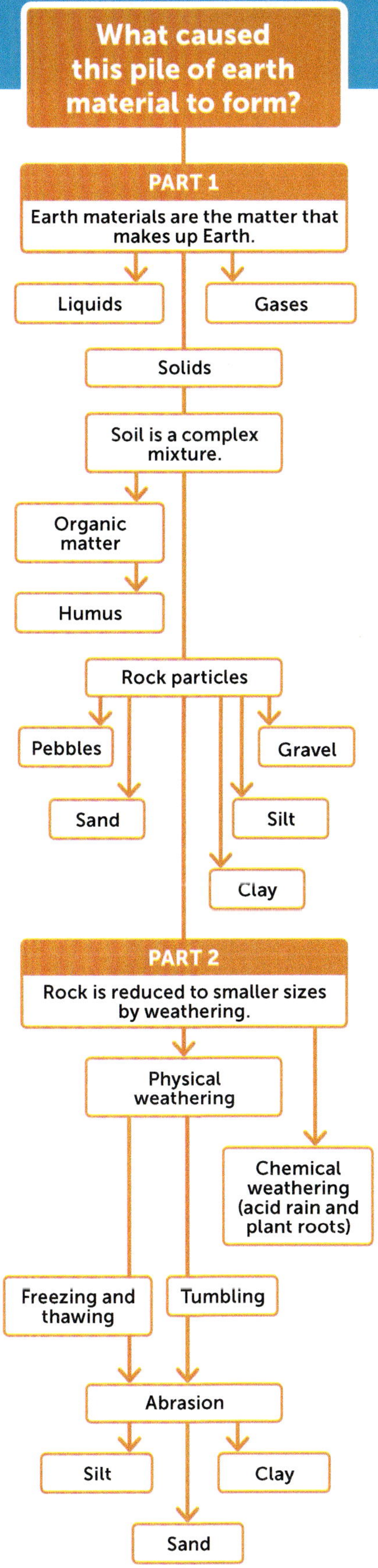

PART 2 Physical Weathering

Materials

PROVIDED EQUIPMENT

For each group

1 *FOSS Science Resources: Soils, Rocks, and Landforms*
"Weathering"
"Making Concrete" (optional)
"Earth Materials in Art" (optional)

1 Basin

1 Plastic jar with cap, 1/2 L

1 Paper plate

For the class

16 Gray granite pieces, tumbled (see Getting Ready)

4 Gray granite pieces, control (never tumbled)

16 Conglomerate pieces, tumbled

4 Conglomerate pieces, control (never tumbled)

2 Zip bags, 1 L

1 Pitcher

SCHOOL-SUPPLIED ITEMS

For the class

1 Glass bottle with tight-fitting cap (See Getting Ready)

- Newspaper

1 Sturdy plastic bag (large enough to contain glass bottle)

- Water
- Permanent marking pen, black
- Safety goggles

FOSSWEB RESOURCES

For each student

- Notebook sheet 3, *Response Sheet—Investigation 1*

For the class

- Word wall cards and equipment cards
- Teaching slides

For the teacher

- *Embedded Assessment Notes*

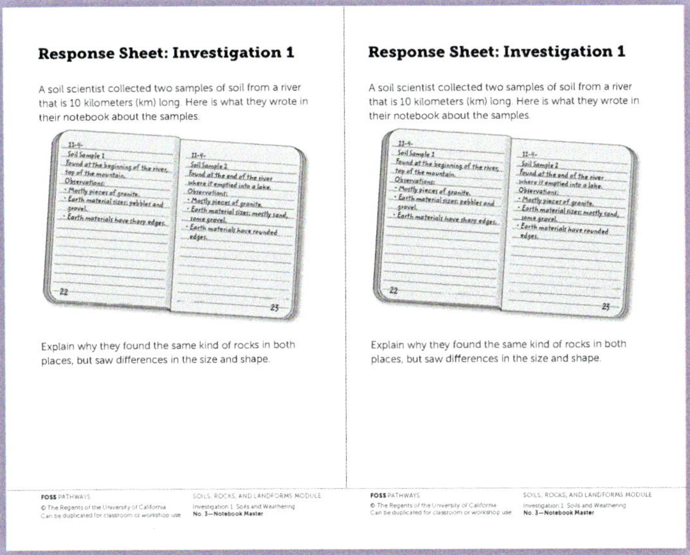

Response Sheet: Investigation 1

A soil scientist collected two samples of soil from a river that is 10 kilometers (km) long. Here is what they wrote in their notebook about the samples.

Explain why they found the same kind of rocks in both places, but saw differences in the size and shape.

FOSS PATHWAYS
© The Regents of the University of California
Can be duplicated for classroom or workshop use

SOILS, ROCKS, AND LANDFORMS MODULE
Investigation 1: Soils and Weathering
No. 3—Notebook Master

Response Sheet: Investigation 1

A soil scientist collected two samples of soil from a river that is 10 kilometers (km) long. Here is what they wrote in their notebook about the samples.

Explain why they found the same kind of rocks in both places, but saw differences in the size and shape.

FOSS PATHWAYS
© The Regents of the University of California
Can be duplicated for classroom or workshop use

SOILS, ROCKS, AND LANDFORMS MODULE
Investigation 1: Soils and Weathering
No. 3—Notebook Master

No. 3—Notebook Master

PART 2 Physical Weathering

Getting Ready

Schedule

4 Sessions (plus 1 optional session)

Session 1: Steps 1–9
Session 2: Steps 10–14
Session 3: Steps 15–18 (Step 18 optional)
Session 4 (optional): Steps 19–20
Session 5: Steps 21–23

Review

- Background for Part 2
- FOSSweb Resources for this part
- Reading: "Weathering"
- Optional readings: "Making Concrete" and "Earth Materials in Art"
- Tutorial: "Weathering" (optional)
- Teacher Preparation Video
- Video: *Freezing Glass Bottle*

Print or Copy

For each student:

- Notebook sheet 3, *Response Sheet—Investigation 1*

For the teacher:

- *Embedded Assessment Notes*

Prepare Material

- Sort granite and conglomerate rock pieces ____ A
- Prepare for the freezing demonstration ____ B

Plan for Assessment

- Review Step 14, "What to Look For" in the response sheet. Response sheets provide formal embedded-assessment data one time in each investigation. Use *Response Sheet—Investigation 1* for a closer look at students' understanding of why the earth materials in different soils have different properties. Plan to spend time having students reflect on their answers after you have reviewed them.

PREPARATION DETAILS

A Sort granite and conglomerate rock pieces

If you are the first person to use this kit, separate the pieces of granite and conglomerate into two groups: rocks that students will tumble (four per group) and control rocks that will not be tumbled (one per group). This allows multiple classes to tumble the rocks, and still have unmarred rocks for comparison. Place the rocks to tumble at the materials station, but hold back the control rocks until Step 6 of Guiding the Investigation.

At the end of the session, put the control rocks into two zip bags that you've labeled "Granite—Control" and "Conglomerate—Control." It is important to keep the control rocks separate from the rocks that are tumbled. Put a note inside the bag letting the next teacher know how many classes have tumbled the rocks.

Shaking the rock pieces in the plastic jars creates more than the normal amount of noise in your classroom. Plan to conduct this part when that noise level will be acceptable.

B Prepare for the freezing demonstration

Gather the materials for the freezing demonstration. Get a glass juice or vinegar bottle (500–1000 mL) with a tight-fitting cap. Narrow-necked bottles work best for this demonstration. The bottle will be filled entirely with water, wrapped in a strong plastic bag, and left overnight in a freezer. Because water expands when it freezes, the bottle will break. The effect is best seen after the water begins to melt and the broken glass falls in pieces.

To prepare, watch the short video, *Freezing Glass Bottle*, on FOSSweb.

NOTE

This is a teacher demonstration. If you are uncomfortable using glass, you could show the video, but it will not be nearly as effective or memorable as a demonstration.

PART 2 Physical Weathering

Guiding the Investigation

SESSION 1 **45 MINUTES**

1 Revisit the earth material on the schoolyard phenomenon

NOTE
The idea here is to help students begin to create a simple model that explains why the rock materials in soils are different from one location to the next.

Have students revisit their thinking about the anchor phenomenon and review any questions or ideas they have about it.

Ask students to think about the soils they observed in the last part. Review the first focus question and students' answers.

➜ *What evidence does the composition of soil provide about where it came from?*

Focus on the rock pieces on the schoolyard. Are all the pieces the same size? Tell students,

Mountains are made of huge masses of rock. The Sierra Nevada, the Appalachian Mountains, and the Rocky Mountains are examples of these huge masses of rock. I'm wondering, if rocks start out as big as mountains, how do they get broken into pieces small enough to be part of the soil?

2 Introduce the focus question

FOCUS QUESTION

What causes big rocks to break into smaller rocks?

Project or write the focus question on the board while you read it aloud.

➜ *What causes big rocks to break into smaller rocks?*

Have students write the date and focus question on the next new page in their science notebook.

3 Discuss the focus question

ENGLISH LANGUAGE SUPPORT
If necessary for student understanding, show pictures that illustrate these actions.

Have students discuss the focus question in their groups for a few moments and then call for responses. Student ideas might include lightning, rocks tumbling down hills, floods, and human activity.

Focus on comments that suggest rocks break apart when they collide with each other. For example, rocks tumble over each other in a river or during a flood, or spill down hills in landslides.

4 Discuss granite and conglomerate rocks

Hold up a piece of **granite** and a piece of **conglomerate**. Add these rock names to the word wall. Ask students to compare and contrast the properties of the granite and the conglomerate. Then ask them to discuss this question.

→ *How could you break these pieces of granite and conglomerate rock into smaller pieces?*

Show students the plastic jars with lids. Ask students to help develop a plan about how to use the jars to see if the rocks can be broken. Write the plan on chart paper if you think students will need help remembering what to do.

a. *Put pieces of granite or conglomerate in a jar. Screw on the lid.*

b. *Shake the jar vigorously for 2 minutes.*

c. *Pour the contents of the jar onto a paper plate and observe.*

MATERIALS FOR STEP 4

- Granite piece, control
- Conglomerate piece, control
- Plastic jar and lid

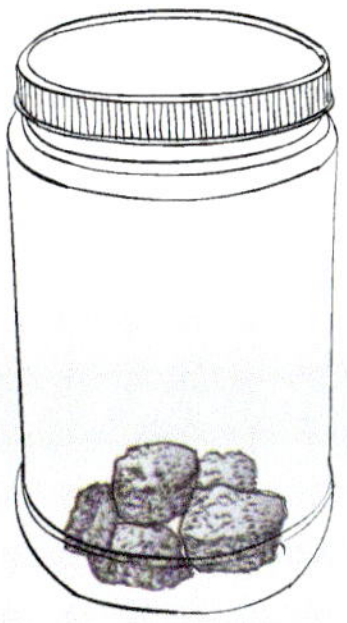

5 Set up and start shaking

Explain that half the groups will tumble granite pieces and the other half will tumble conglomerate pieces. Have Getters go to the materials station and use a basin to pick up one jar with a lid, four pieces of one kind of rock—granite or conglomerate, as assigned—one paper plate, and four pairs of safety goggles. Plan on groups shaking for about 2 minutes total, with each student getting a turn. Students can count to 30 to approximate 30 seconds for a turn. Have students carry out the investigations.

MATERIALS FOR STEP 5

- Basins
- Plastic jars
- Pieces of granite or conglomerate, tumbling
- Paper plates
- Safety goggles

6 Observe effect of tumbling

Caution students to let the dust settle a bit before opening the jars, then pour the contents of the jar onto a paper plate, being careful not to breathe in the dust. Have students observe and record the effects of tumbling on a new page in their notebooks. Have them include labeled drawings of the rocks. Be sure they include all the different sizes of rock pieces they see, including the dust particles.

Give each group a fresh untumbled control rock to compare with those that have been tumbled. Have students compare the tumbled rocks with the control piece of rock and draw a labeled picture of the untumbled rock

Students should note that nothing changed except the size and shape of the pieces of rock. They ended up with the same kind of rocks, just smaller pieces.

Science and Engineering Practices

- Planning and carrying out investigations

Crosscutting Concepts

- Cause and effect

PART 2 Physical Weathering

SESSION 1

7 Trade rocks

Have groups that tumbled granite switch paper plates and rock materials with a group that tumbled conglomerate. Students should label and draw the same observations for the second kind of rock, then compare the two kinds of rocks.

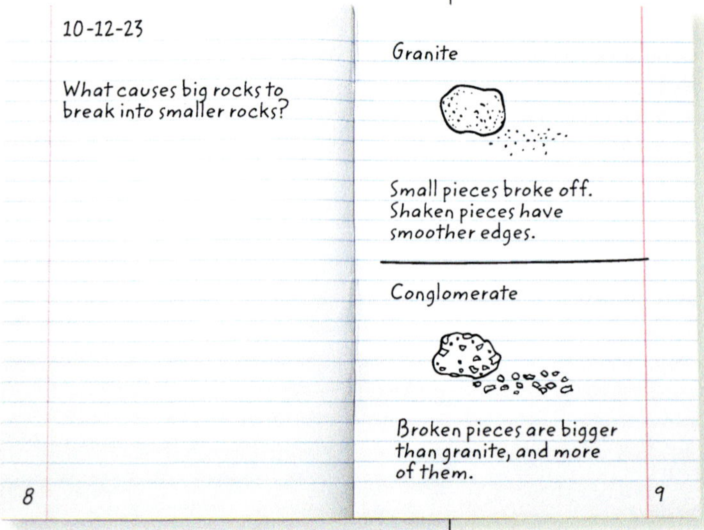

Ask students to describe what happened to the rocks (the effect) that were tumbled (the cause). Ask,

→ *What is the effect of tumbling the rocks?* [Small sand-size pieces broke off the granite. The conglomerate broke into more and bigger pieces than the granite. The tumbled rocks are more rounded and have smoother edges than the control rocks.]

Have Getters return the jars, safety goggles, and rock materials to the materials station. Return rock pieces to the appropriate storage bags, making sure to keep the control rocks in separate bags.

Crosscutting Concepts
- Cause and effect

ENGLISH LANGUAGE SUPPORT
To help with comprehension, make a simple diagram as you explain and label it with the words in bold.

8 Have a sense-making discussion

Ask students to discuss these questions with a partner and then to share their ideas and build on the ideas of others in a whole group discussion.

→ *What is the effect of tumbling (cause) on the rocks?* [Break into smaller pieces.]

→ *What happens to the pieces when they break off?* [They might be carried to another place by wind or water and settle.]

→ *How does this explain the pieces of rock in the soil?* [This is how soil forms, from small pieces of rock settling.]

Tell students,

All rocks, including mountains, exposed at Earth's surface break apart over time. Some rocks break apart faster than others. ***Weathering*** *is the word scientists use to describe the breakdown of rock into smaller pieces.*

When rocks break into smaller pieces without changing what the rock is made of, it is called ***physical weathering****. One kind of physical weathering is* ***abrasion*** *(rubbing together). Abrasion occurs when rocks fall, when rocks tumble in landslides, and when rocks hit one another as they are pushed around by moving water, waves, and wind.*

Add the new vocabulary to the class word wall. As a class, make a concept definition map for physical weathering. Start with the definition of physical weathering in the What is it? box. Then review what the characteristics are in the What is it like? boxes [rocks breaking apart; rocks rubbing together].

Next add some examples [rocks falling in landslides, rocks hitting one another in moving water, waves knocking rocks around; wind].

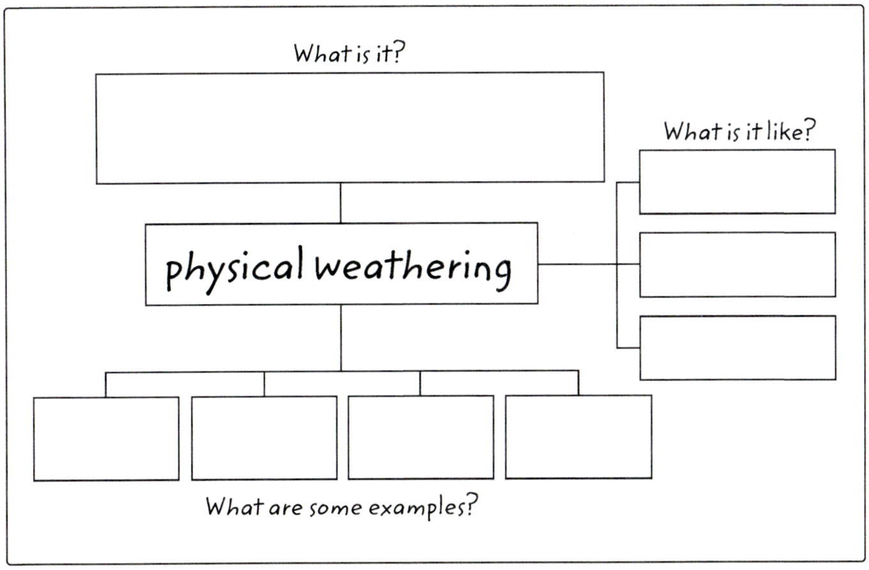

ELA CONNECTION

This suggested strategy addresses this English Language Arts objective.

Demonstrate understanding of word relationships and nuances in word meanings.

9 Set up the freezing bottle demonstration

Hold up a glass bottle and tell students,

Water sometimes flows into cracks in rocks. When the temperature gets below 0°C, water will ***freeze****.*

Let's pretend this bottle is a rock. I'm going to fill it with water. We'll put it into the freezer today and take a look at it tomorrow.

Pause for students to discuss with a partner what they think will happen.

This bottle-water system will serve as a ***model*** *of what might happen in the natural world. A model is a representation of the features and actions of a natural system or process, but it isn't exactly the same as the real world—it focuses on the most important features. Scientists use models to help them observe and explain what happens in systems that move too slowly or too quickly, or ones that are too big or too small to observe directly.*

Fill the bottle completely (absolutely no air space), and screw the cap on tightly. Place the bottle in a strong plastic bag (to contain the glass pieces when the bottle breaks) and roll it up tightly. Place the bagged bottle in a freezer overnight.

MATERIALS FOR STEP 9

- Glass bottle
- Sturdy plastic bag
- Newspaper
- Pitcher of water
- Freezer

Science and Engineering Practices

- Developing and using models

PART 2 Physical Weathering

SESSION 2 45 MINUTES

10 Observe the frozen bottle

Retrieve the bottle from the freezer. Carefully open the plastic bag and put it in a location where the students can observe the results from a distance. Cracks should be visible in the bottle. Let it sit out for about an hour so it begins to thaw. As it thaws, the glass pieces may begin to break off from the ice. Students should not be near or touch the broken glass.

Ask students to offer explanations for the fractured glass using a cause-and-effect statement such as "I think the (cause) made the (effect)."

After students have offered their ideas, tell them,

When water freezes, it ***expands****. That means the volume of the water increases—it gets bigger. When it expands, the water pushes on things around it with a lot of force. The force was enough to break this bottle.*

In the mountains and in other places where it is cold in winter, rocks are exposed to cycles of freezing and thawing.

→ *When the water in a crack in a rock freezes, what might happen to the rock?* [Freezing water can break a rock into smaller pieces.]

This cycle of freezing and thawing—water flowing into cracks when it's liquid and expanding when it freezes—is another kind of physical weathering that breaks rock into smaller pieces.

Carefully transfer the ice and broken glass to a basin. After the ice melts, pour off the water and recycle the glass.

NOTE
Ask students to also wear safety goggles while observing the demonstration.

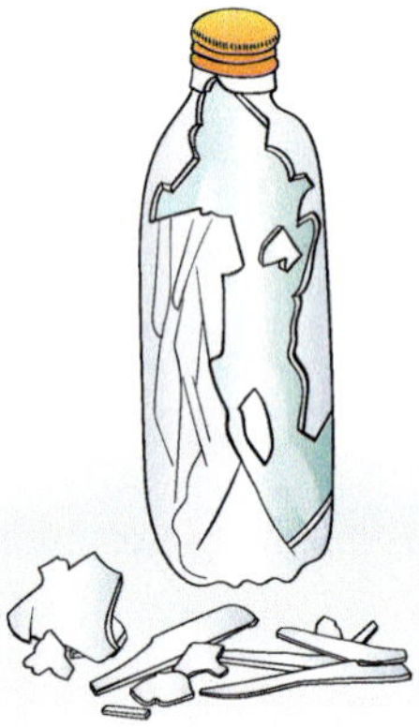

ENGLISH LANGUAGE SUPPORT
Make a "before and after" diagram of the bottle of water and together with students add words and arrows to explain what happened when it froze.

11 Connect to the anchor phenomenon

Ask students what aspects of the weathering of rocks might help explain the appearance of the properties of the earth material that appeared at the bottom of the hill.

You might ask,

→ *How does this explain the size of the rocks in the pile of earth material?*

→ *What other ideas do you have for how rocks might break into smaller pieces?*

→ *What new questions do you have?*

NOTE
The video *Freezing Glass Bottle* on FOSSweb can be used for this demonstration, but seeing it firsthand is more engaging.

Crosscutting Concepts
- Cause and effect

12 Review vocabulary

Review key vocabulary introduced in this part. If these words are not on the word wall, add them now. Add *freezing* and *thawing* to the concept definition map as another example of physical weathering.

abrasion
conglomerate
expand
freeze
granite
model
physical weathering
weathering

13 Respond to the focus question

Ask students to work independently to construct an explanation about one way rocks break down into smaller pieces.

→ *What causes big rocks to break into smaller rocks?*

Ask students to think back to the effects of what they did with the granite and the conglomerate rocks in the shaking jar and to consider the action of the frozen water in the glass bottle. Ask,

→ *How do the results of those investigations help us to understand what happens in the natural world with really big rocks and mountains?*

For students who need scaffolding, provide a sentence frame such as: *Big rocks break down into smaller pieces by ______. We found out that ______.*

Science and Engineering Practices

- Constructing explanations

Crosscutting Concepts

- Cause and effect

14 Assess progress: response sheet

Distribute notebook sheet 3, *Response Sheet—Investigation 1*. Have students glue the response sheet to the next blank left-hand page, and write their responses on the right-hand page.

Ask students to hand in their notebooks open to this page. Review their responses after class for a closer look at students' understanding of why the earth materials in different soils have different properties, and use the data to prepare for the next lesson.

WHAT TO LOOK FOR

- *Students analyze and interpret the data in the notebook to state that both locations have the same kind of rock because the river moves the rock from mountaintop to the lake.*
- *Students explain that more abrasion has occurred as the rocks were moved down the river. This effect caused the rocks to be smaller and smoother near the lake.*

Use this next-step strategy if you determine one is needed. See the Assessment chapter for more on Next-Step Strategies.

Use the key points and revision with color strategy. Begin by discussing the information on the response sheet. After it is clear that students understand what is intended, call on individuals or groups to suggest key points that should be included in a complete answer. Write the key points on the board or chart paper as phrases or individual words that will scaffold students' revision.

Science and Engineering Practices

- Analyzing and interpreting data

Disciplinary Core Ideas

- ESS2.A: Earth materials and systems

Crosscutting Concepts

- Cause and effect

SESSION 2

Guide students to discuss and chart these important key points:

- Both locations have the same kind of rock because the river is moving the same source rock. The difference is location on the river.
- Rocks are smaller farther down the river because more abrasion occurs the farther the rock travels.
- Rocks are more rounded farther down the river because the more abrasion that occurs, the more rough edges are smoothed.

After the class discussion, use colored pens or pencils and the three Cs (confirm, complete, and correct) to revise their answers. As students read over their responses, they confirm correct information by underlining with a green pen; they complete their responses by adding information that is missing using a blue pen; and they correct wrong information using a red pen.

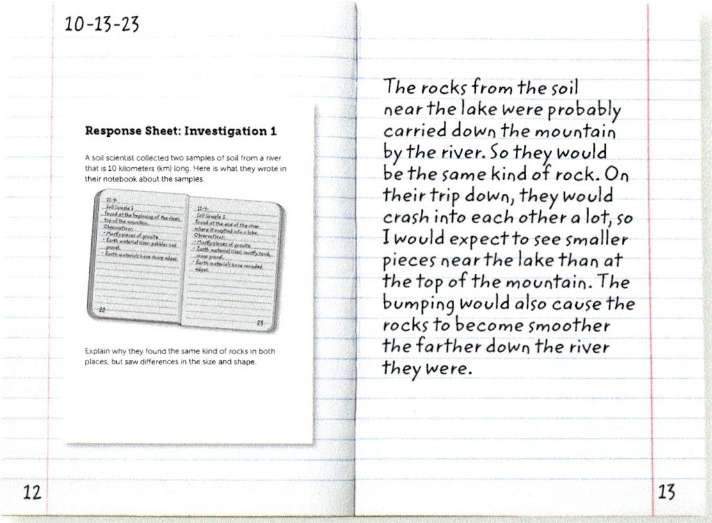

Interpreting Information

SESSION 3 **45 MINUTES**

15 Read "Weathering"

Have students read "Weathering" to obtain and communicate information about the effects of weathering on earth materials. The article helps students connect what they have investigated in the classroom about physical and chemical weathering to the landforms in the natural world that are affected by these processes.

16 Use a photo preview and predict strategy

Before reading, give students a few minutes to look at the photographs. Tell them to choose a few photos to discuss with a reading partner. Have them take turns explaining what they think is happening in the photos. If needed, provide sentence frames such as:

- *I think ______ because ______.*
- *I agree/disagree with you because ______.*

Read the article aloud or have students read it independently to look for evidence that supports the ideas they discussed.

Pause after each page to match the information in the text with the photographs. Have students discuss with their partner how the photos help them understand the different ways rocks can be broken down into smaller pieces.

Science and Engineering Practices

- Obtaining, evaluating, and communicating information

ELA CONNECTION

This suggested strategy addresses this English Language Arts objective.

Interpret information presented visually, and explain how the information contributes to an understanding of the text.

PART 2 Physical Weathering

SESSION 3

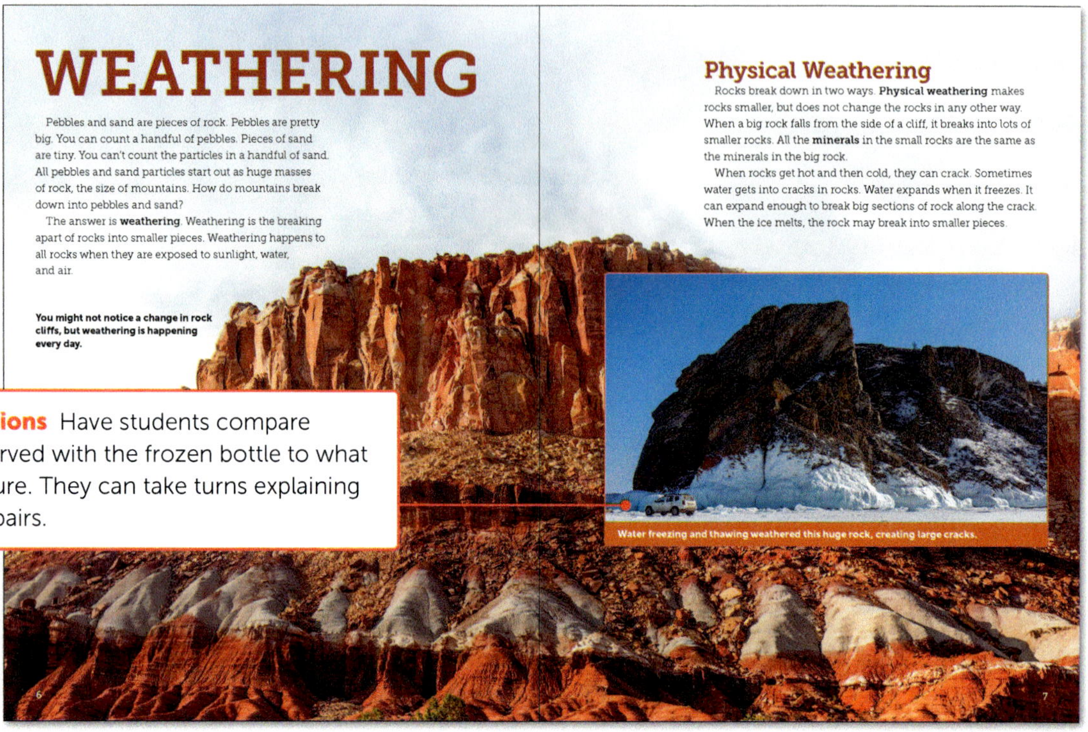

WEATHERING

Pebbles and sand are pieces of rock. Pebbles are pretty big. You can count a handful of pebbles. Pieces of sand are tiny. You can't count the particles in a handful of sand. All pebbles and sand particles start out as huge masses of rock, the size of mountains. How do mountains break down into pebbles and sand?

The answer is **weathering**. Weathering is the breaking apart of rocks into smaller pieces. Weathering happens to all rocks when they are exposed to sunlight, water, and air.

You might not notice a change in rock cliffs, but weathering is happening every day.

Physical Weathering

Rocks break down in two ways. **Physical weathering** makes rocks smaller, but does not change the rocks in any other way. When a big rock falls from the side of a cliff, it breaks into lots of smaller rocks. All the **minerals** in the small rocks are the same as the minerals in the big rock.

When rocks get hot and then cold, they can crack. Sometimes water gets into cracks in rocks. Water expands when it freezes. It can expand enough to break big sections of rock along the crack. When the ice melts, the rock may break into smaller pieces.

Water freezing and thawing weathered this huge rock, creating large cracks.

6

7

Make Connections Have students compare what they observed with the frozen bottle to what happens in nature. They can take turns explaining the process in pairs.

Roots of trees and bushes can grow down into cracks in rocks. As roots grow, they make the cracks bigger. Sometimes the cracks get so big that the rock splits apart.

When rocks move against one another, they get worn down. Rubbing, grinding, and tumbling are forms of **abrasion**. Abrasion is a kind of physical weathering. It happens when rocks fall in **landslides**, tumble in flowing water, or roll around in waves. Wind can blow sand against rocks. This sandblasting weathers the rocks.

Tree roots grow into cracks and break rocks.

Sand carried by wind can weather rocks into interesting shapes.

Chemical Weathering

Chemical weathering happens when minerals in rocks are changed by chemicals in water and air. The minerals change into new substances.

Many rocks contain iron. When oxygen in air comes in contact with iron, the iron in the rock can rust. Rust is iron oxide. Iron oxide is softer than other iron minerals. This causes the rock to break apart faster.

Chemical weathering of a rock containing iron

8

9

Activate Prior Knowledge Ask students to share prior observations they have made of plants growing in the cracks of rocks or other forms of weathering.

Carbon dioxide gas in the air **dissolves** in water droplets. This forms an **acid**. The acid droplets can fall as rain. The acid dissolves the **calcite** in **limestone** and **marble** to form holes. This is a chemical change. Statues, buildings, and gravestones made of marble or limestone change when exposed to acid rain.

Salt can cause chemical weathering. Salt water can **react** with minerals in rocks to make new minerals. When the new substances are softer than the original mineral, holes can form. The weakened rock breaks and falls apart more easily.

Ask Questions Have students share questions they have about chemical weathering that they would like to investigate or research.

Chemical weathering of sandstone by salt water

Chemical weathering of marble by acid rain

Questions for Reflection

1. What is the effect of acid rain on rocks?
2. How do living organisms contribute to the weathering of rocks?
3. How does ice cause weathering?

10

PART 2 Physical Weathering

Crosscutting Concepts

- Cause and effect

ELA CONNECTION

These suggested strategies address these English Language Arts objectives.

Explain procedures, ideas, or concepts in a scientific text.

Determine the meaning of general academic domain-specific words or phrases.

Describe the overall structure of information in a text.

Refer to details and examples in a text when explaining what the text says explicitly and when drawing inferences from text.

Engage in collaborative discussions.

SESSION 3

17 Discuss the reading

Review any unknown words or phrases. Ask students what they notice about the structure of the text [cause/effect] and how thinking about cause and effect helps them understand weathering. Review the language structures for cause and effect, e.g., When... then... . Have students read the article a second time, looking for cause-and-effect statements.

Ask students to discuss the difference between physical and chemical weathering with a partner based on the information in the text. You might have students use this strategy in pairs to help them listen closely and express their ideas clearly.

First round:

1. *Partner A:* Describe physical weathering to your partner. (Partner B is silent.)
2. *Partner B:* Repeat what your partner said.
3. *Partner A:* Confirm or correct what your partner repeated.

Students should respond that physical weathering breaks rock into smaller pieces without changing the nature of the rock itself.

Second round:

1. *Partner B:* Describe chemical weathering to your partner. (Partner A is silent.)
2. *Partner A:* Repeat what your partner said.
3. *Partner B:* Confirm or correct what your partner repeated.

Students should respond that chemical weathering breaks down rock by changing minerals into new minerals.

Continue the discussion using these questions to check for understanding. Encourage students to cite details and examples from the text to support their answers.

→ *What is the effect of acid rain on rocks?* [Acid reacts with the ingredient calcite in limestone and marble to cause weathering. This weakens the rock.]

→ *How do living organisms contribute to the weathering of rocks?* [Plants send roots (living things) into cracks in rocks. As roots grow, they break rocks apart to cause physical weathering; humans contribute to acid rain.]

→ *How does ice cause weathering?* [Water expands when it freezes. Water in cracks in rocks acts like a wedge to break rocks apart.]

Conclude by having students connect the ideas in the article to what they have investigated in class. Ask,

- *What observations have you made that support the idea of physical weathering in the natural world?*
- *What is different about the investigations in class and how things happen in the natural world?*

Encourage students to cite evidence from their notebooks as well as the article to support their ideas.

ELA CONNECTION

This suggested strategy addresses this English Language Arts objective.

Compare and contrast a firsthand and secondhand account of the same topic.

18 Engage with the tutorial: "Weathering" (optional)

This tutorial serves as a good review of physical and chemical weathering and can be viewed before or after the reading of the same name.

PART 2 Physical Weathering

SESSION 4 **45 MINUTES (OPTIONAL)**

ELA CONNECTION

These suggested reading strategies address these English Language Arts objectives.

Refer to details and examples in a text when explaining what the text says explicitly and when drawing inferences from text.

Conduct short research projects that build knowledge through investigation of different aspects of a topic.

Engage in collaborative discussions.

Report on a text in an organized manner, using appropriate facts and relevant, descriptive details to support main ideas or themes.

19 Read "Making Concrete"

Have students read the article "Making Concrete." This article gives students a look at concrete production, from large-scale projects to their own schoolyard.

Before starting the reading, give students a few minutes to preview the text by looking at the photographs and captions. Tell students to choose one photograph and share with a partner what they think is happening and any personal connections they may be able to make. Ask students to think about what information they will obtain from reading this article.

Read the article aloud, or have students read independently.

After the reading, have students share new information they learned from the reading with their table groups. Review any unfamiliar words, phrases, or ideas students might have. Discuss the reasons why local aggregates are used to make concrete [to make the concrete cost effective].

Use the Questions for Reflection to check for understanding. Encourage students to cite evidence from the text to support their ideas.

➜ *Why did scientists and engineers study the concrete used to make the Roman Colosseum? What did they discover?* [To find out what the concrete was made of because it lasted for thousands of years. They found the concrete was made with materials different from modern-day cement. It contained volcanic ash.]

➜ *Why is this important to us today?* [Engineers want to design concrete structures to be stronger and last longer.]

Ask students to re-read the last paragraph and discuss the question at the end.

➜ *Where did the aggregates come from to make the concrete foundation used to build your school?*

Ask students how they might go about finding out the answer to this last question. This question could lead to a short research project students conduct to find out more about how natural resources are used in their local community.

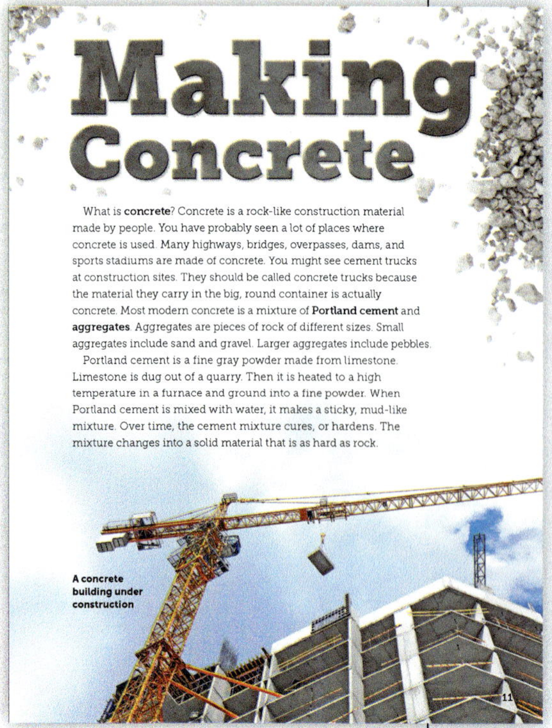

Making Concrete

What is **concrete**? Concrete is a rock-like construction material made by people. You have probably seen a lot of places where concrete is used. Many highways, bridges, overpasses, dams, and sports stadiums are made of concrete. You might see cement trucks at construction sites. They should be called concrete trucks because the material they carry in the big, round container is actually concrete. Most modern concrete is a mixture of **Portland cement** and **aggregates**. Aggregates are pieces of rock of different sizes. Small aggregates include sand and gravel. Larger aggregates include pebbles.

Portland cement is a fine gray powder made from limestone. Limestone is dug out of a quarry. Then it is heated to a high temperature in a furnace and ground into a fine powder. When Portland cement is mixed with water, it makes a sticky, mud-like mixture. Over time, the cement mixture cures, or hardens. The mixture changes into a solid material that is as hard as rock.

A concrete building under construction

11

20 Read "Earth Materials in Art"

This article explains how an artist uses an earth material, clay, to make tiles as pieces of art. Start by having students focus on the photograph of the woman cutting the clay. Have students share what they can see in the photograph. Then have them share what they infer. Help generate responses by asking,

- *What can you assume based on what you see?*
- *What can you conclude about what is happening or will happen?*

Tell students that this article will give them more information about the artist in the photograph and how she makes clay tiles.

Read the article aloud or have students read independently or in guided reading groups.

After the first read, have students read the article a second time and jot down notes in their notebooks. Use the 3-2-1 strategy to help students with comprehension. Have them write 3 important details; 2 interesting things; and 1 question.

Give students a few minutes to compare their notes with their group. Then, call for a few volunteers to share their questions. Note the questions on chart paper for possible future discussion or research topics. Have students take turns describing the process of making tiles out of clay. Encourage them to use the photographs as they retell the process. Ask students what other earth materials artists might use.

Focus students' attention on the text structures of the article. Ask,

- *What structures did the author use to organize the text?* [Sequence, cause and effect.]
- *What are some examples?*

ELA CONNECTION

These suggested reading strategies address these English Language Arts objectives.

Determine the main idea of a text.

Describe the overall structure of information in a text.

Interpret information presented visually, and explain how the information contributes to an understanding of the text.

Gather relevant information from experiences and print.

Earth Materials

in Art

Rose Craig (1943–) is an artist. Craig makes beautiful watercolor paintings, and she is a skilled graphic artist and illustrator. Craig has another artistic interest too. Craig makes ceramic tiles. Ceramic tiles are made of clay. Because clay is an earth material, she is an earth material artist.

This is how she makes her beautiful tiles. First, she places a big chunk of clay on her worktable. She uses a big rolling pin to flatten the clay into one big, thin sheet. The rolling process is very similar to rolling pie dough for a pie crust. When the clay is rolled out just right, Craig uses a straightedge and a knife to cut the slab of clay into rectangular pieces. Craig then uses rubber stamps to press a shape into the surface of the soft clay. She might use a fish shape, a flower shape, or a dragonfly shape. Sometimes she trims around the shape. Then Craig waits for the clay to dry.

First, the clay is cut. Then it is rolled out flat, and stamped with a shape.

17

PART 2 Physical Weathering

Reflecting on a Phenomenon

SESSION 5 **45 MINUTES**

21 Revisit the partial explanation of the anchor phenomenon

Ask students to think about the different soils on the schoolyard that they have observed.

➜ *Why might there be differences in the soil in different locations?* [Different conditions and amounts of rock particles and dead plants available to become parts of the soil.]

➜ *If the new pile of earth material and the hill soil in the scenario are similar, what might that tell us?* [The pile of earth material came from the hill.]

➜ *If they were very different, what could that tell us?* [The pile of earth material didn't come from the hill.]

Have students add to their existing explanation.

Students should start to make the connection that some of the earth material in the soil are similar to those in the pile, but there are some differences. This will be explored more in Investigation 2.

Ask students some questions to set them up for erosion and deposition in the next investigation.

➜ *What might have caused the soil to appear at the bottom of the hill?*

➜ *What information would you like to collect to inform your responses?*

➜ *What new questions do you have?*

22 Share notebook entries

Conclude Part 2 by having students share notebook entries.

→ *What causes big rocks to break into smaller rocks?*

Ask students to pair up with a partner to:

- compare their answers to the focus question;
- share their answers to the response sheet.

Discuss cause-and-effect relationships. Point out that science is about trying to find out the reasons why things happen and developing models to illustrate the process. Ask students to think about what causes rocks to break into smaller pieces [abrasion, freezing]. With a partner, have them discuss what effects these produce. Encourage students to provide evidence from the investigation and from their own experiences. Here are some sentence frames students can use to build on each other's ideas.

- *So what you're saying is ______.*
- *What caused ______ to happen?*
- *What was the effect of ______?*
- *What is your evidence that ______?*
- *How else could you find evidence for ______?*

Another instructional strategy to engage students in argumentation is to provide a statement such as "Freezing causes most of the physical weathering in (location)." Students agree or disagree with that statement and provide some reasoned thoughts about how this might work in the real world.

ELA CONNECTION

This suggested strategy addresses this English Language Arts objective.

Engage in collaborative discussions.

ENGLISH LANGUAGE SUPPORT

See the Science-Centered Language Development in Grades 3–5 chapter for strategies on sharing notebook entries.

Science and Engineering Practices

- Developing and using models
- Engaging in argument from evidence

Crosscutting Concepts

- Cause and effect

NOTE

See the **Home/School Connection** for Investigation 1 at the end of the Differentiation Opportunities section. This is a good time to send it home with students.

INVESTIGATION 1

PART 2 Physical Weathering

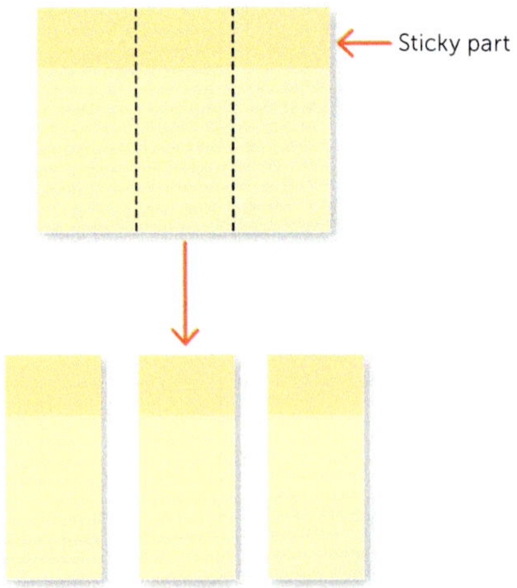

THREE-DIMENSIONAL LEARNING OBJECTIVES

- Students observe, record, and analyze data to reveal patterns about the different materials found in soil. (Part 1)
- Students use patterns from firsthand observations and media to engage in argument about the source of different types of soil. (Part 1)
- Students plan and carry out investigations to determine the effects of weathering on different earth materials such as those found near schoolyards. (Part 2)

SESSION 5

23 Review three-dimensions for Investigation 1

Distribute one or two self-stick notes to each student. Ask students to cut each note into three pieces, making sure that each piece has a sticky end.

Ask students to take a few minutes to look back through their notebook entries to find the most important things they learned in Investigation 1. Students should include at least one science and engineering practice, one disciplinary core idea, and one crosscutting concept. They should tag those pages with self-stick notes.

When students have completed their notebook review and placed their tags, lead a short class discussion to create a list of three-dimensional statements that summarize what students have learned in this investigation.

By using science and engineering practices and exploring core ideas through the lens of crosscutting concepts, students should come forward with these big ideas in the review discussions.

- We conducted investigations to learn about the properties of soils from different places.
- From our data analysis, there is a pattern in soil composition—soils are composed of small rocks (gravel, sand, silt, or clay) and humus.
- We modeled different kinds of physical weathering: abrasion and freezing and thawing.
- We read gathering information (reading) about chemical weathering and compared it to physical weathering.

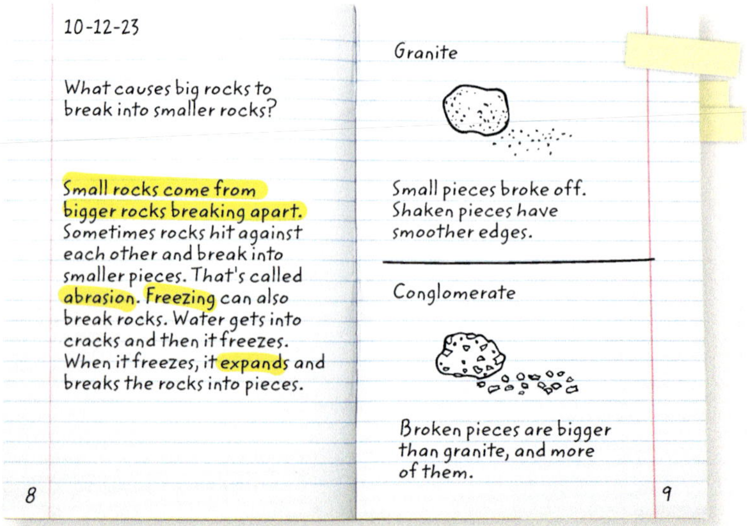

SIDE TRIP 1

Schoolyard Soils

Focus on environmental literacy

Purpose

This is an optional activity, a side trip from the main pathway, that provides an extended experience in environmental literacy that can be used with the whole class, or it can be used for a student-choice challenge, or for differentiation.

- The detailed lesson plan is a printable digital resource found on FOSSweb including notebook master 4, *Schoolyard Soil Samples*.

FOCUS QUESTION

What's in our schoolyard soils?

Activity Summary

Students collect and observe different soil samples from several locations on the schoolyard. They analyze the soil samples to determine how much humus and rock material are in local soils. Students can use this information to add to their understanding of the anchor phenomenon.

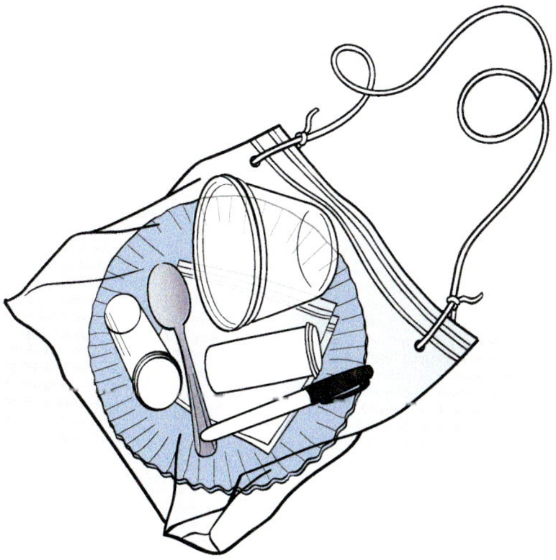

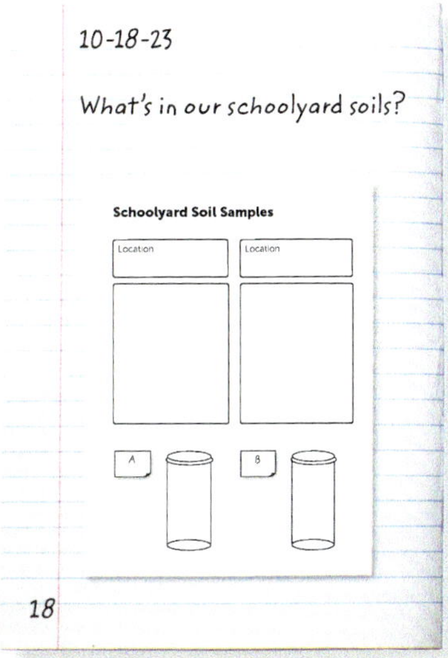

Science and Engineering Practices

- Planning and carrying out investigations
- Analyzing and interpreting data

Disciplinary Core Ideas

- ESS2.A: Earth materials and systems

Crosscutting Concepts

- Patterns

PART 2 Physical Weathering

SIDE TRIP 2

Chemical Weathering

Focus on carrying out investigations

Purpose

This is an optional activity, a side trip from the main pathway, that provides an extended experience in carrying out investigations to test cause-and-effect relationships that can be used with the whole class, or it can be used for a student-choice challenge, or for differentiation.

- The detailed lesson plan is a printable digital resource found on FOSSweb including notebook masters 5, *Rock Observations*, 6, *Rocks in Acid Rain*, 7, *Acid-Rain Evaporation*, and teacher master 2, *Reference Rocks for Acid-Rain*.
- The video *Chemical Weathering*, available on FOSSweb, is a resource for this side trip.

FOCUS QUESTION

How are rocks affected by acid rain?

Activity Summary

Students plan and conduct an investigation to test rocks for interaction with "acid rain" which is one form of chemical weathering. They see that some rocks (limestone and marble) are very susceptible to acid rain, but other rocks (basalt and sandstone) are unaffected.

Science and Engineering Practices

- Developing and using models
- Planning and carrying out investigations
- Analyzing and interpreting data
- Constructing explanations

Disciplinary Core Ideas

- ESS2.A: Earth materials and systems
- ESS2.E: Biogeology

Crosscutting Concepts

- Cause and effect

Differentiation Opportunities

Below is a sample of student-choice activities related to this investigation for group or individual projects. Students can select an activity that is of special interest to them, or the class can choose a project for collaboration. Extensions and project ideas are written for students and accessible on FOSSweb.

Language Extension

- Research organisms that live in soils and write about a day in the life of those organisms

Science Extensions

- Look for local soil profiles at roadside cuts
- Compare local soil samples
- Engage with the tutorial: "Weathering"

Art Extensions

- Read the article "Earth Materials in Art" in *FOSS Science Resources* book
- Collect soils to make fingerpaints to create a piece of art

Environmental Literacy Extensions

- Find out what information is available from the U.S. Geological Survey or the local Soil Conservation Agency about local soils
- Read the article "Making Concrete" in *FOSS Science Resources* book

Home/School Connection

Students are encouraged to find and test five or six different rocks from their neighborhood for interaction with acid rain (vinegar). Students explain to family members that acid rain interacts with a mineral called calcite, an ingredient in many rocks. Together they test each rock to observe what happens and what that tells them about the presence of calcite in each rock.

Print or make copies of *Home/School Connection* for Investigation 1, and send it home with students after Part 2.

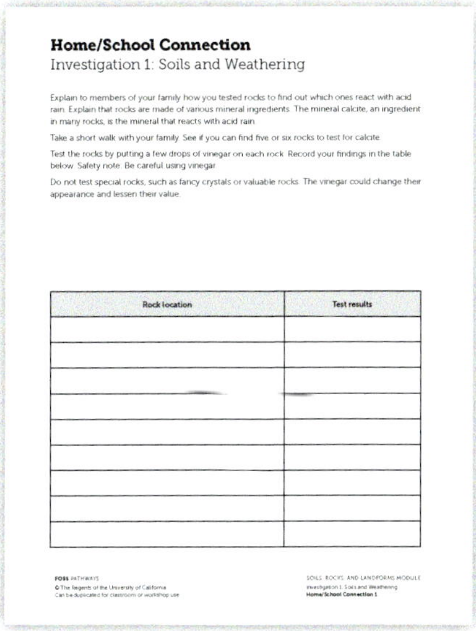

Home/School Connection
Investigation 1: Soils and Weathering

Explain to members of your family how you tested rocks to find out which ones react with acid rain. Explain that rocks are made of various mineral ingredients. The mineral calcite, an ingredient in many rocks, is the mineral that reacts with acid rain.

Take a short walk with your family. See if you can find five or six rocks to test for calcite.

Test the rocks by putting a few drops of vinegar on each rock. Record your findings in the table below. Safety note: Be careful using vinegar.

Do not test special rocks, such as fancy crystals or valuable rocks. The vinegar could change their appearance and lessen their value.

Rock location	Test results

FOSS PATHWAYS
© The Regents of the University of California
Can be duplicated for classroom or workshop use

SOILS, ROCKS, AND LANDFORMS MODULE
Investigation 1: Soils and Weathering
Home/School Connection 1

No. 1—Home/School Connection

INVESTIGATION 2

Landforms

INVESTIGATION 2

Landforms

Phenomenon Storyline

The phenomenon storyline from Investigation 1 is continued in this investigation. On a walk around the schoolyard, students discover some changes at the edge of the playground. They observe a new pile of sand and dirt near the bottom of a small hill.

What caused this pile of earth material to form?

In Investigation 1, students gathered data to determine the composition of soils from different locations and discovered that not all soils are the same. They investigated the process of weathering, which break down rocks to form soils. In this investigation, students create a model stream-table system to investigate the variables involved with the phenomena of erosion and deposition of weathered earth material due to flowing water. Finally, they engage in argument from evidence to determine where the sand and dirt came from and how it was deposited there.

After using a physical stream table as a model system to investigate what happens in the natural world, students work with a virtual model system to gather data to compare the impact of different engineering designs to reduce the impact of erosion and landslides on human structures.

THREE-DIMENSIONAL LEARNING OBJECTIVES

- Students plan and carry out investigations using physical stream-table models to determine the effects of erosion and deposition on landforms. **(Parts 1–2)**
- Students observe, record, and analyze data to determine cause-and-effect relationships about the rate and volume of erosion and the energy of moving water or wind. **(Part 2)**
- Students use a virtual modeling tool to develop possible solutions that reduce the impact of hazards from erosion and gather data to make evidence-based engineering decisions about the stability and change of human and natural systems. **(Part 3)**

INVESTIGATION 2

Landforms
At a Glance

Investigation Part	Sessions	Practices and Crosscutting Concepts
Continue anchor phenomenon 1—Pile of earth material on the schoolyard		
PART 1 Erosion and Deposition **FOCUS QUESTION: What causes weathered rock pieces to move from one place to another?** Students use stream tables to investigate that water moves earth materials from one location to another. After running a volume of water through the stream table, students shake a vial containing a sample of earth material mixed with water to observe the rate at which different-sized particles of earth material settle out. These observations provide information about the origin of earth materials in the schoolyard.	**SESSION 1*** Steps 1–11 **SESSION 2** Steps 12–16 **SESSION 3** Steps 17–22 *A session is 45 minutes.	**Practices** • Developing and using models • Planning and carrying out investigations • Analyzing and interpreting data • Constructing explanations • Obtaining, evaluating, and communicating information **Crosscutting Concepts** • Cause and effect • Scale, proportion, and quantity • Systems and system models
PART 2 Stream-Table Investigations **FOCUS QUESTIONS: How does slope affect erosion and deposition? How do floods affect erosion and deposition?** Students continue to run stream tables to learn how environmental variables can affect erosion and deposition. They investigate the variables of slope and water volume (flood) to further understand how the pile of earth material formed on the schoolyard.	**SESSION 1** Steps 1–2 **SESSION 2** Steps 3–12 (Step 12 optional) **SESSION 3** Steps 13–19 (Step 19 optional) **SESSION 4** Steps 20–24 (Steps 22, 24 optional)	**Practices** • Developing and using models • Planning and carrying out investigations • Analyzing and interpreting data • Constructing explanations • Engaging in argument from evidence • Obtaining, evaluating, and communicating information **Crosscutting Concepts** • Cause and effect • Scale, proportion, and quantity • Systems and system models • Energy and matter
PART 3 Reducing Erosion Impact **FOCUS QUESTION: What solutions can we design to reduce the impact of erosion and deposition on human structures?** Students use a digital simulation tool to ask questions, plan investigations, and gather data to make evidence-based decisions about how to mitigate the impact of weathering, erosion, and deposition on natural and human structures.	**SESSION 1** Steps 1–12 **SESSION 2** Steps 13–16 **SESSION 3** Steps 17–20 **SESSION 4** Steps 21–22 *I-Check 1–2*	**Practices** • Asking questions and defining problems • Developing and using models • Planning and carrying out investigations • Analyzing and interpreting data • Constructing explanations and designing solutions **Crosscutting Concepts** • Cause and effect • Systems and system models • Stability and change

Performance Expectations

4-ESS2-1. Make observations and/or measurements to provide evidence of the effects of weathering or the rate of erosion by water, ice, wind, or vegetation.

4-ESS3-2. Generate and compare multiple solutions to reduce the impacts of natural Earth processes on humans.

3–5-ETS1-2. Generate and compare multiple possible solutions to a problem based on how well each is likely to meet the criteria and constraints of the problem.

Disciplinary Core Ideas	Interpreting Information	Assessment
ESS2.A: Earth materials and systems • Weathered rock material can be reshaped into new landforms by the processes of erosion and deposition. • Erosion is the transport (movement) of weathered rock material (sediment) by moving water or wind. • Deposition is the settling of sediment when the speed of moving water or wind declines.	**Science Resource Book** "Erosion and Deposition" **Other Media** Video: *Erosion and Deposition* (optional) Video: *Sediments Time-Lapse*	**Embedded Assessment** Science notebook entry **NGSS Performance Expectation addressed** 4-ESS2-1
ESS2.A: Earth materials and systems • The rate and volume of erosion relate directly to the energy of moving water or wind. • The energy of moving water depends on the mass of water in motion and its velocity. The greater the mass and velocity, the greater the energy.	**Science Resource Book** "Landforms on Display" "Where Do Rocks Come From?" "Monumental Rocks" (optional) **Other Media** Video: *Stream-Table Investigations* (optional) Tutorial: "Soil Formation" (optional)	**Embedded Assessment** Response sheet **NGSS Performance Expectation addressed** 4-ESS2-1
ESS2.A: Earth materials and systems **ESS2.E:** Biogeology **ESS3.B:** Natural hazards **ETS1.B:** Developing possible solutions • Erosion is the transport (movement) of weathered rock material (sediments) by moving water or wind. • Deposition is the settling of sediments when the speed of moving water or wind declines. • Scale models can be used to gather data and make decisions about natural systems.	**Other Media** Online Simulation: "Erosion Engineering"	**Embedded Assessment** Performance assessment **Benchmark Assessment** *Investigations 1–2 I-Check* **NGSS Performance Expectations addressed** 4-ESS2-1 4-ESS3-2 3–5-ETS1-2

INVESTIGATION 2

PART 1 Erosion and Deposition

Science and Engineering Practices

- Developing and using models
- Planning and carrying out investigations
- Analyzing and interpreting data
- Constructing explanations
- Obtaining, evaluating, and communicating information

Disciplinary Core Ideas

- ESS2.A: Earth materials and systems

Crosscutting Concepts

- Cause and effect
- Systems and system models
- Scale, proportion, and quantity

Aa

alluvial fan
canyon
delta
deposition
erosion
landform
river mouth
sediment
valley

Teacher Background

Phenomenon Storyline

LEARNING OBJECTIVE:

- **Students plan and carry out investigations using physical stream-table models to determine the effects of erosion and deposition on landforms.**

Parts 1 and 2 revisit the anchor phenomenon of the pile of earth material in the schoolyard. In Investigation 1, students observed the effects of weathering. They now continue to build an explanation by adding on the effect of erosion and deposition on landforms. Here are some things they will find out to connect to the storyline.

- Weathered pieces of rock can be moved from place to place by water and wind (erosion). Water flowing downhill carries the rock pieces to new locations.
- Weathered rock particles settle out to form piles of earth materials (on a small scale) and to form new landforms (on a large scale). The settling out is called deposition.

FOCUS QUESTION

What Causes Weathered Rock Pieces to Move from One Place to Another?

Earth is a dynamic place. Great forces continually reshape the face of the planet. Some of these forces bring molten material to the surface or bend and raise the planet's crust, forming new mountains or ridges and valleys. Opposing forces, such as moving water and wind, wear down the land and fill in irregularities on the planet's surface. If Earth's crust weren't continually being built up, the planet might end up as smooth as a bowling ball.

We observe the phenomenon of erosion as the transportation of earth materials from one location to another. Water, working with the force of gravity, is the most powerful agent of erosion on Earth. Flowing water tumbles and transports pieces of rock. Tumbling rocks break into smaller pieces, while simultaneously acting as powerful abrasives, wearing away the rock over which they move. The general rule is that the farther a rock travels, the more it collides with other rocks and the smaller it becomes. This is how boulders are reduced to silt. In addition to physical weathering, chemical weathering processes are usually involved in the transformation of large rock particles into minute clay particles.

Wind picks up small rock particles and transports them from one place to another. Ice in the form of glaciers also picks up and moves rocks, while grinding them into smaller particles. After thousands of years of erosion, mountains become rounded, and river channels deepen and widen. In some places, curious landforms, such as hoodoos (eroded pillars or columns of rock in bizarre shapes) and hogbacks (sharp, eroded ridges of steeply sloped rocks), decorate the landscape.

Rock particles are classified by size with specific diameters for each type of particle. Boulders are large; cobbles are intermediate in size; pebbles are small; gravel is yet smaller bits of rock; sand is tiny rock particles; silt particles require a hand lens to be seen; and clay particles are microscopic.

There is a reciprocal phenomenon to erosion. Earth materials, carried by water, wind, and ice, eventually settle out in a process called deposition. The deposited rock particles are called sediment. Sediment can consist of huge boulders or motes of silt and clay.

A powerful flood might move a boulder a few meters in a riverbed. Sand, silt, and clay may remain suspended in water for long distances. Particles travel with flowing water until the velocity of the current decreases enough for the particles to yield to the pull of gravity and settle on the bottom as sediment. This kind of deposition characteristically occurs where a stream or river pours out into slow-moving or still water in a lake or the ocean, resulting in a landform called a delta at the river mouth. Alluvial fans are similar to deltas, except they form on land where the slope of the terrain decreases at the lower end of a steep canyon. The amount of erosion of earth materials depends to a large extent on the amount of energy available in the moving water. The greater the mass and speed of water, the greater the energy of the water, and the greater the energy, the larger the particles that can be put into motion, and the farther those particles will travel.

Smaller rock particles (sand, silt, and clay) can also be transported by wind. Large deposits of wind-blown sediment form sand dunes in desert and coastal areas.

The conceptual flow for this investigation starts with a model of a simulated landform, a plain composed of a mixture of earth material (sand and clay) packed into one end of a stream table (plastic tray). Water enters the system and flows across the surface of the earth material plain. As water flows, students observe that the water carves a canyon across the plain. The canyon continues to grow as the running water erodes sand and clay, carrying it into the lower portion of the tray where it is deposited. The erosion and deposition create new landforms, including a river valley across the plain and a delta or alluvial fan where the sand and clay accumulate in the lower reaches of the tray.

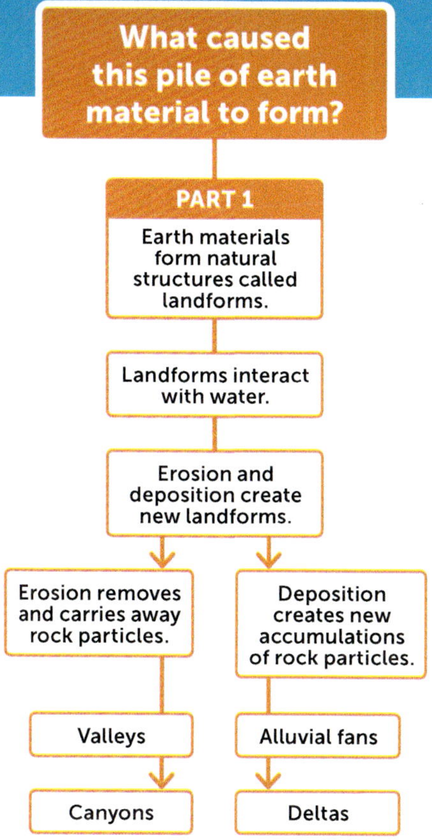

INVESTIGATION 2

PART 1 Erosion and Deposition

NOTE
The sand/clay mixture is permanent equipment. Dry it thoroughly after the investigation before returning it to the kit. These materials come in a gray plastic box and should be stored in the same box when the entire stream-table investigation is finished.

Materials

PROVIDED EQUIPMENT

For each student

- **1** *FOSS Science Resources: Soils, Rocks, and Landforms* "Erosion and Deposition"

For each group

- **1** Stream table
- **1** Wood angle
- **1** Basin
- **1** Vial with cap
- **1** Standard water source, 1/2 L (blue dot)
- **1** Ruler, 30 cm
- **1** Container, 1 L (without lid)
- **1** Meter tape
- **1** Syringe

For the class

- **2** Plastic cups
- **2** Bags of white, powdered clay (don't use gray powdered clay)
- **8** Bags of fine sand
- • Duct tape
- **1** Pitcher

SCHOOL-SUPPLIED ITEMS

For the class

- Water
- Newspaper
- Safety goggles, gloves, and face masks for dust (optional)
- Paper towels
- **1** Dustpan and broom

FOSSWEB RESOURCES

For each student

- Notebook sheet 8, *Standard Stream-Table Setup*
- Notebook sheet 9, *Stream-Table Observations*
- Notebook sheet 10, *Landform Vocabulary*

For the class

- Teacher master 3, *Standard Stream-Table Setup*
- Word wall cards and equipment cards
- Teaching slides
- Anchor Phenomenon Resource Video 1

For the teacher

- *Embedded Assessment Notes*

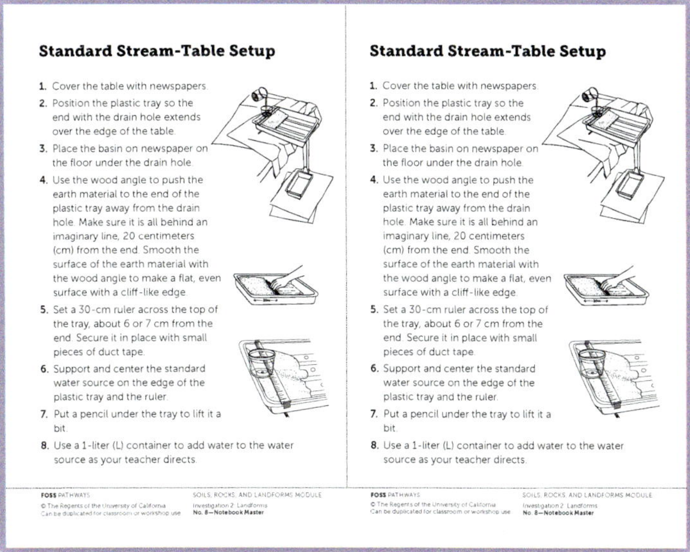
Standard Stream-Table Setup

1. Cover the table with newspapers.
2. Position the plastic tray so the end with the drain hole extends over the edge of the table.
3. Place the basin on newspaper on the floor under the drain hole.
4. Use the wood angle to push the earth material to the end of the plastic tray away from the drain hole. Make sure it is all behind an imaginary line, 20 centimeters (cm) from the end. Smooth the surface of the earth material with the wood angle to make a flat, even surface with a cliff-like edge.
5. Set a 30-cm ruler across the top of the tray, about 6 or 7 cm from the end. Secure it in place with small pieces of duct tape.
6. Support and center the standard water source on the edge of the plastic tray and the ruler.
7. Put a pencil under the tray to lift it a bit.
8. Use a 1-liter (L) container to add water to the water source as your teacher directs.

FOSS PATHWAYS
© The Regents of the University of California
Can be duplicated for classroom or workshop use.
SOILS, ROCKS, AND LANDFORMS MODULE
Investigation 2: Landforms
No. 8—Notebook Master

No. 8—Notebook Master

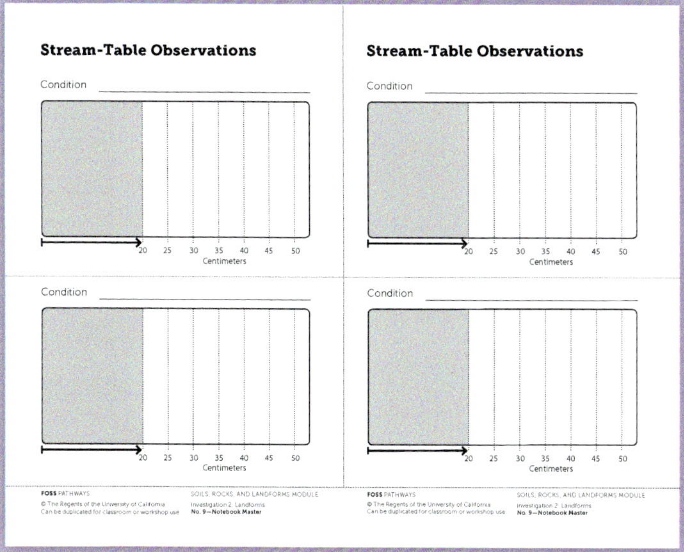
Stream-Table Observations

Condition ____

Condition ____

FOSS PATHWAYS
© The Regents of the University of California
Can be duplicated for classroom or workshop use.
SOILS, ROCKS, AND LANDFORMS MODULE
Investigation 2: Landforms
No. 9—Notebook Master

No. 9—Notebook Master

Landform Vocabulary

FOSS PATHWAYS
© The Regents of the University of California
Can be duplicated for classroom or workshop use.
SOILS, ROCKS, AND LANDFORMS MODULE
Investigation 2: Landforms
No. 10—Notebook Master

No. 10—Notebook Master

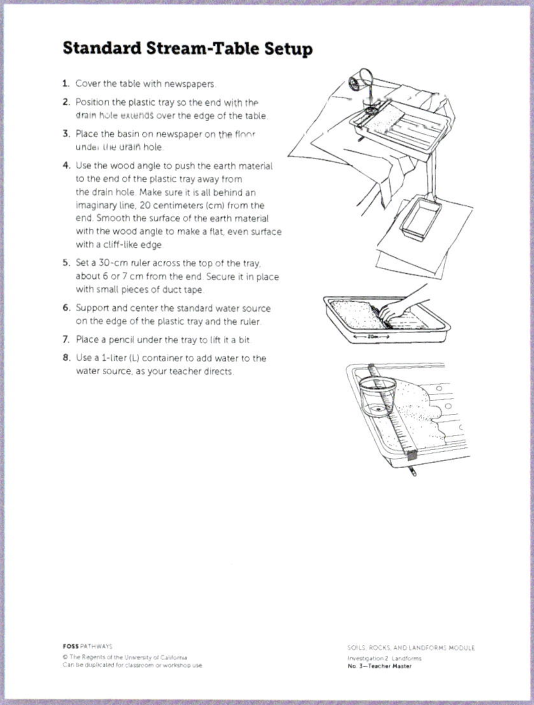
Standard Stream-Table Setup

1. Cover the table with newspapers.
2. Position the plastic tray so the end with the drain hole extends over the edge of the table.
3. Place the basin on newspaper on the floor under the drain hole.
4. Use the wood angle to push the earth material to the end of the plastic tray away from the drain hole. Make sure it is all behind an imaginary line, 20 centimeters (cm) from the end. Smooth the surface of the earth material with the wood angle to make a flat, even surface with a cliff-like edge.
5. Set a 30-cm ruler across the top of the tray, about 6 or 7 cm from the end. Secure it in place with small pieces of duct tape.
6. Support and center the standard water source on the edge of the plastic tray and the ruler.
7. Place a pencil under the tray to lift it a bit.
8. Use a 1-liter (L) container to add water to the water source, as your teacher directs.

FOSS PATHWAYS
© The Regents of the University of California
Can be duplicated for classroom or workshop use.
SOILS, ROCKS, AND LANDFORMS MODULE
Investigation 2: Landforms
No. 3—Teacher Master

No. 3—Teacher Master

INVESTIGATION 2

PART 1 Erosion and Deposition

Getting Ready

Schedule

3 Sessions
Session 1: Steps 1–11
Session 2: Steps 12–16
Session 3: Steps 17–22

Review

- Background for Part 1
- Materials chapter for stream-table setup, storage, cleanup, and safety
- FOSSweb Resources for this part
- Reading: "Erosion and Deposition"
- Video: *Erosion and Deposition* (optional)
- Video: *Sediments Time-Lapse*
- Teacher Preparation Video

Print or Copy

For each student:

- Notebook sheet 8, *Standard Stream-Table Setup*
- Notebook sheet 9, *Stream-Table Observations*
- Notebook sheet 10, *Landform Vocabulary*

For the teacher:

- *Embedded Assessment Notes*

Prepare Material

- Prepare earth material for stream table ______ A
- Practice using stream table ______ B
- Prepare the work area ______ C

Plan for Assessment

- Review Step 16, "What to Look For" in the notebook entry

PREPARATION DETAILS

A Prepare earth material for stream table

Students flow water over a mixture of sand and clay (earth material) in a stream table to observe erosion and deposition. If you are the first person to use the kit, you will need to mix the sand and clay. The sand and clay used for the stream table is different from that used to make the soils in Investigation 1. Look for the stream-table earth materials in the small gray plastic box that comes as part of the kit.

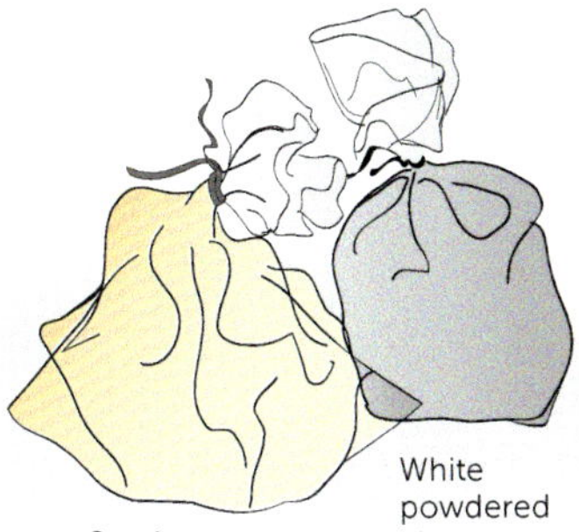

Before you mix them, set aside one plastic cup of sand and another of clay. You will use these to show students the separate materials in Step 4.

These instructions will make a batch of earth material for one tray. Enlist the help of students, parents, or aides to mix the number of batches needed for your class.

a. Add about 100 grams (1 plastic cupful) of white powdered clay to a 1.35 kilogram (3 pounds, about 1 liter) bag of fine sand.

b. Stir the clay and sand until they are uniformly mixed.

c. Add up to 1 cup of water to the dry materials in the bag and mix thoroughly.

If the earth material has been used previously, you may need to add a little sand or clay to each bag. The exact amounts are not critical. There should be more sand than clay in the mixture, and the total volume of earth material should be about 1 liter. You might also need to break up the mixture if it has hardened from the last use.

B Practice using stream table

Before beginning this investigation, set one up, as described on notebook sheet 8 and teacher master 3, *Standard Stream-Table Setup*. Conduct several trials, noting the landforms that develop and the texture of the earth material after several uses.

C Prepare the work area

It is important to have level surfaces for the stream tables. Plan to put newspaper on desktops and under the catch basins. Some earth material may flow through the drain hole and into the catch basin. After it settles, pour off the water and return the earth material to the stream table. Don't dump the earth material down a drain. Dumping the basins outdoors where there is soil is a better option. Provide paper towels and a broom and dustpan to help clean up spills. If you are concerned about getting water and earth material on the floor, consider conducting the investigation outdoors.

PART 1 Erosion and Deposition

Guiding the Investigation

ENGLISH LANGUAGE SUPPORT
Use the labeled diagrams from the previous investigation to review these ideas.

Science and Engineering Practices
- Asking questions

Science and Engineering Practices
- Developing and using models

Crosscutting Concepts
- Systems and system models

SESSION 1 **45 MINUTES**

1 Revisit the pile of earth materials phenomenon

Revisit the phenomenon from Investigation 1 using the Anchor Phenomenon Resource Video 1. Ask students to discuss the processes of weathering and how it provides information to explain how the earth material formed. To guide the discussion, you might ask,

➜ *What is weathering?* [The breaking of rock into smaller pieces.]

➜ *What was in the soil that might explain where the earth material came from?* [Small pieces of rock that are similar to those on the schoolyard.]

Ask students what new questions they have about the pile of earth materials. When students communicate their curiosity about how the pieces of materials move from the soil to the schoolyard, introduce the focus question.

2 Introduce the focus question

FOCUS QUESTION

What causes weathered rock pieces to move from one place to another?

Ask students the focus question as you project or write it on the board, and have students copy it into their notebooks with the date.

➜ *What causes weathered rock pieces to move from one place to another?*

3 Introduce *model*

Explain that one way geologists investigate the movement of earth materials is to develop and study models. Review what a model is.

A model is a representation of the features and actions of a natural system or process, but it isn't exactly the same as the real world—it focuses on the most important features. Scientists use models to help them observe what happens in systems that move too slowly or too quickly or are too big or too small to observe directly.

4 Describe stream tables

Show students a stream table. Ask students how they could use this model to find out more about where the earth materials came from. Tell them,

This stream table is a model of a flat piece of land. In this model, the land is composed of soil made of two particle sizes: sand and clay.

Show students the samples of sand and clay in two plastic cups.

Give each student a copy of notebook sheet 8, *Standard Stream-Table Setup*. Project the *Standard Stream-Table Setup* teacher master as you explain the procedure for preparing and using the stream table.

1. Cover the table with newspapers.
2. Position the plastic tray so the end with the drain hole extends over the edge of the table.
3. Place the basin on newspaper on the floor under the drain hole.
4. Use the wood angle to push the earth material to the end of the plastic tray away from the drain hole. Make sure it is all behind an imaginary line, 20 centimeters (cm) from the end. Smooth the surface of the earth material with the wood angle to make a flat, even surface with a cliff-like edge.
5. Set a 30-cm ruler across the top of the tray, about 6 or 7 cm from the end. Secure it in place with small pieces of duct tape.
6. Support and center the standard water source, on the edge of the plastic tray and the ruler.
7. Place a pencil under the tray to lift it a bit.
8. Use a 1-liter (L) container to add water to the water source, as your teacher directs.

Water-source container 1 L container

MATERIALS FOR STEP 4

- Stream table
- **Standard Stream-Table Setup** sheet
- Cup of sand
- Cup of clay

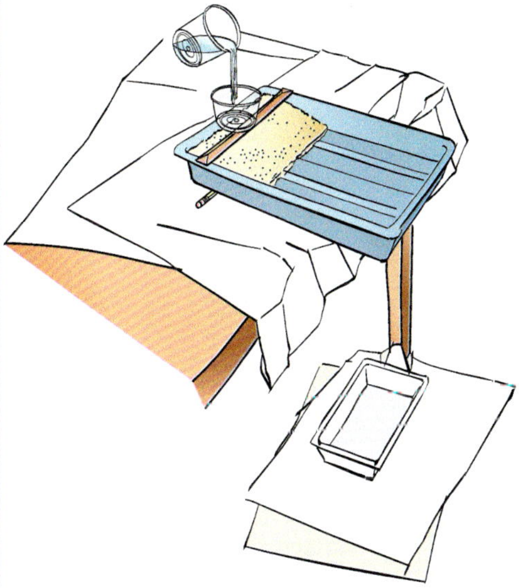

INVESTIGATION 2

PART 1 Erosion and Deposition

MATERIALS FOR STEP 5

- Stream tables
- Bags of earth material
- Standard water sources
- Rulers
- Wood angles
- Containers, 1 L
- Basins
- Newspapers
- Duct tape
- Water
- Meter tapes
- **Stream-Table Observations** sheets
- Safety goggles and gloves (optional)

NOTE

Provide safety goggles if required by your district's guidelines. You may also want to provide gloves to students with sensitive skin.

Science and Engineering Practices

- Planning and carrying out investigations

SESSION 1

5 Set up stream tables

Ask Getters to pick up materials for their groups. Have the Starters supervise the stream-table setup. As groups complete the stream-table setups, give them notebook sheet 9, *Stream-Table Observations*, to glue into their notebooks. Students should label the condition for the first stream table on the sheet "standard—before run" and the second stream table on the sheet "standard—after run." They should also draw in the stream-table setup—before run.

Students can take pictures of the stream tables after setup and after running the water through. Students can print the photos and glue them into their notebooks, label the photos, and write notes to go with the pictures. This can be done in addition to the notebook sheet 9 maps students draw or as an alternative.

6 Carry out the investigation

Remind students that everyone will run exactly 1 L of water through the stream tables. Explain that it is important for everyone to use the same amount of water since this is one of the variables to be controlled in later runs. While the water is flowing, no one should touch the stream-table tray or the earth material in the tray or bump or shake the table.

If another class previously set up the stream tables, students may need to remove duct tape from the hole in the stream tables before using the water.

Send Getters to fill the 1 L containers with water. Let them begin the investigation by pouring the liter of water into the standard water-source container positioned over the earth material in the stream table.

7 Analyze and interpret data

When all the water has run through the stream-table models hold a 2-minute discussion to review what students observed. Ask them to think about these questions.

- ➜ *Where did the earth material start in the tray, and where did it go? the sand? the clay?*
- ➜ *How does this model system show what happens to earth materials in nature?*
- ➜ *What does this tell us about the effect of flowing water on earth materials?*

Have students draw a second picture of the stream table on notebook sheet 9, showing how it looks after the water has run to reveal patterns.

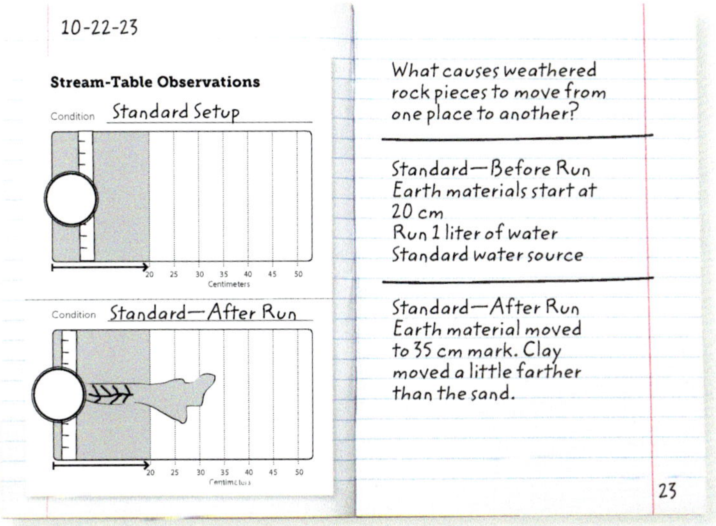

8 Share observations and build vocabulary

Have the Reporter from each group report one or two of the effects of the water on the earth materials. As students report the various formations they have seen, develop landform vocabulary.

For example, when students mention the depression in the earth through which the stream of water flowed, identify that space as a **valley** or a **canyon**. When they mention the fan-shaped materials deposited at the stream or **river mouth**, identify it as an **alluvial fan** or **delta**.

NOTE

Go to FOSSweb for the Science and Engineering Practices—Grade 4 chapter for details on how to engage students with the practice of developing and using models.

Crosscutting Concepts

- Patterns
- Cause and effect

Science and Engineering Practices

- Developing and using models
- Analyzing and interpreting data

PART 1 Erosion and Deposition

SESSION 1

9 Distribute the *Landform Vocabulary* sheets

Have Getters pick up copies of notebook sheet 10, *Landform Vocabulary*, for their groups. Have students glue this sheet onto the left-hand side of a two-page spread. Leave the right-hand page blank for definitions that will be added later. Project the sheet to help students identify the valley and write the name in the appropriate location.

Have students label the canyon and delta. Explain that the **landform** that was created in the stream tables is called an *alluvial fan* because it formed on "dry land." When this landform is created in a body of water, it is called a delta. The landform shown on the sheet is a delta. Ask,

➜ *What did you notice when you looked at the results of other groups' stream-table investigations?*

10 Introduce *erosion* and *deposition*

As students report their observations, connect their ideas to erosion and deposition. Say,

*One thing you observed in your stream tables is called **erosion**. Erosion is the movement of pieces of weathered rock by water or wind. Erosion changes the shape of the land. In your stream tables, water eroded the earth material forming a valley or a canyon. A valley is the broad area through which a river flows. A canyon is a narrow, very steep-sided valley.*

*Earth material carried by moving water eventually stops moving and settles out. The settling process is called **deposition**. The eroded materials that settle are called **sediment**. When the water slows down, sediment is deposited in a new location, somewhere along the path of the moving water. The fan-shaped deposits of sediment you see at the mouth of your stream-table river form an alluvial fan.*

Emphasize that *erosion* means to take away (from), and *deposition* means to place or settle out. Thus, eroded materials end up being deposited materials.

ENGLISH LANGUAGE SUPPORT
Make a simple diagram of the stream table and label it with the words in bold and other words students may need as a reference. Use arrows to show the movement of earth material.

NOTE
This is important to emphasize. Many students think erosion and deposition are the same thing.

11 Prepare equipment for next session

Distribute a vial and lid to each group. Have them fill the vial half full with earth material from the stream table and put the lid on the vial. The vials will be used in the next session.

Have students place a small piece of duct tape over the drain hole in their trays and stack the trays crisscross in an appropriate location.

Have students return everything but the stream table with the earth material to the materials station. Ask the Starters to recycle the water in the basins outdoors and clean the tabletops.

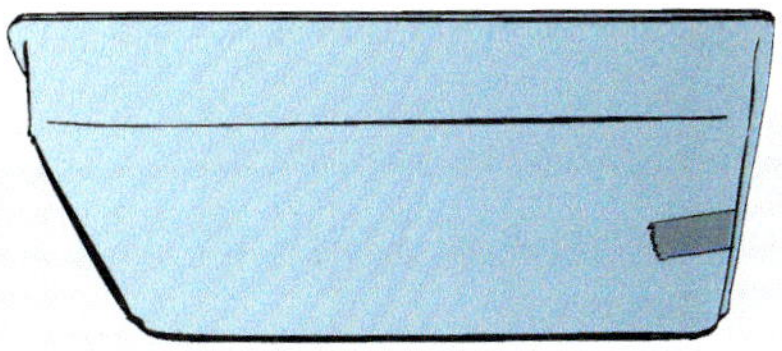

NOTE
Allow the basins of water to settle after running the stream tables. Decant most of the water. Blot off as much water as possible. Stir the clay back into the stream-table earth material.

PART 1 Erosion and Deposition

MATERIALS FOR STEP 12
- Vials with caps
- Syringes
- Water

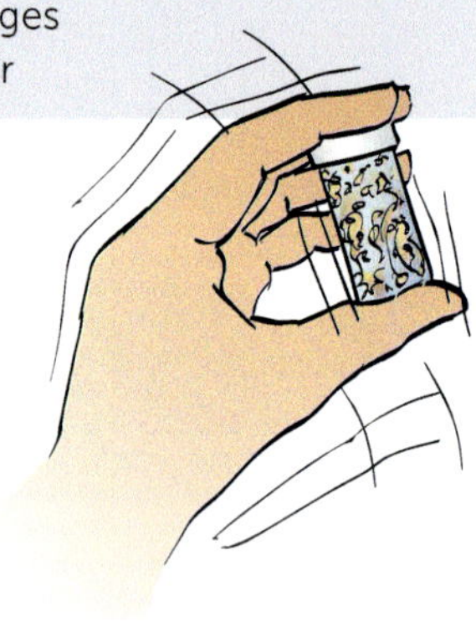

SESSION 2 45 MINUTES

12 Conduct the shake test

Remind students that many of them noticed that some of the eroded earth materials were carried farther by the flow of water than others. Ask,

→ *Why do you think the clay was deposited farther downstream?*

Have students conduct the earth-materials-shake test (like they did in Investigation 1, Part 1, with the soils) to give them additional information that will help them think about why some eroded materials travel farther than others.

Give each group a 12-dram vial with cap that they prepared with stream-table earth material the day before. Provide these instructions.

a. *Use the syringe to transfer water into the vial until it is almost full.*

b. *Put the cap on the vial. Shake it for 5 seconds.*

c. *Set the vial on the table and observe for a minute or two.*

Let students conduct the shake test. Remind students that their task is to observe the rate at which different materials settle. Students will notice that the larger-grained sand particles settle to the bottom of the vial as soon as it is placed on the table. The smaller clay particles settle more slowly, forming a distinct clay layer on top of the sand.

13 Have a sense-making discussion and connect to the anchor phenomenon

Science and Engineering Practices
- Developing and using models
- Constructing explanations

Crosscutting Concepts
- Cause and effect
- Systems and system models

The video *Sediments Time-Lapse* is a good visual to use for guiding students to construct explanations about deposition during the sense-making discussion.

Ask,

→ *What was the effect when you shook the earth material with water?* [The sand immediately settled to the bottom of the vial; clay settled on top of the sand.]

→ *Why do you think the sand was deposited on the bottom of the vial with the clay on top of the sand?* [Sand is heavier (larger particles) and settled to the bottom first when the shaking stopped.]

→ *What caused the sand and clay to deposit in different places in the stream-table system model?* [Larger (heavier) earth materials settled as soon as the water flow started to slow down. Smaller (lighter) earth materials, such as clay particles, stayed suspended in the water longer, so they traveled farther and didn't settle until the flow stopped.]

- → *How can you relate this to soils? How was the settling in the stream table like that in the soil vials? How might the size of the rock particles in soils at different locations be determined by erosion and deposition? Explain your ideas.* [The settling process was just like that in the soil vials. Smaller particles stayed suspended longer which is why they are deposited farther downstream.]
- → *How could this help you explain where the pile of earth materials in the schoolyard scenario came from?*

14 Review vocabulary

Review the vocabulary words that have been introduced. Make sure that these words are on the word wall. Add any other words students might need.

15 Respond to the focus question

Have students consider the focus question and write a response in their notebooks.

- → *What causes weathered rock pieces to move from one place to another?*

Encourage students to use their stream-table investigations and new vocabulary words in their answers.

16 Assess progress: notebook entry

Have students hand in their notebooks open to the page on which they answered the focus question. After class, review students' entries to see if students clearly understand the difference between *erosion* and *deposition*.

WHAT TO LOOK FOR

- *Students use their observations of the stream-table results to explain when the water was poured on the earth materials, some of it was carried away (eroded) and formed canyons or valleys.*
- *Students explain that some earth materials were deposited in another place (alluvial fan or delta).*
- *Students explain how the stream-table systems model the process of land formations in nature (modeling).*

Aa

alluvial fan
canyon
delta
deposition
erosion
landform
river mouth
sediment
valley

ENGLISH LANGUAGE SUPPORT

For students who need scaffolding, provide a sentence frame such as:

Weathered rock pieces move from one place to another by _____.
In our stream table we noticed _____.

Science and Engineering Practices

- Developing and using models
- Constructing explanations

Disciplinary Core Ideas

- ESS2.A: Earth materials and systems

Crosscutting Concepts

- Cause and effect
- Systems and system models

Interpreting Information

SESSION 3 **45 MINUTES**

17 Read "Erosion and Deposition"

Have students read the article "Erosion and Deposition" to obtain, evaluate, and communicate information about the processes that move and deposit earth material, creating landforms.

Before reading, give students time to preview the text by reading the subtitles, looking at and discussing the photographs and captions, and reviewing the words in bold. Review that the stream-table investigation is one way to present a model of erosion and deposition. Explain that this article will give them more information about how these processes work in the natural world.

18 Use a summarizing reading strategy

Tell students to read each section independently and to think about how their observations in class compare to the ideas presented in the text. To support reading comprehension, have students pause after each section and then take turns summarizing with a partner. Model by reading and summarizing the first page aloud.

Have students read the next section on erosion. Pause and have Partner A give the main idea; Partner B gives the details. Repeat the process for deposition starting with Partner B. Have students read the last section independently and then take turns summarizing.

Science and Engineering Practices

- Obtaining, evaluating, and communicating information

ELA CONNECTION

These suggested strategies address these English Language Arts objectives.

Compare and contrast a firsthand and secondhand account of the same topic.

Interpret information presented visually, and explain how the information contributes to an understanding of the text.

Determine the main idea of a text and explain how it is supported by key details; summarize the text.

Erosion and Deposition

A trip to the beach is fun. One of the best parts is playing in the sand. Where did all that sand come from? Was it made right there, or did it come from some other place?

Much of the sand on the beach came from mountains. **Erosion** moved the sand from the mountains to the beach. Erosion is the moving away of weathered rock. After rocks have weathered into small pieces, they can be carried away by gravity, water, or wind. Most of the sand shown here was carried to the beach by water flowing in rivers and streams.

As long as river water keeps flowing, the bits of sand keep moving downstream. When the river reaches the ocean, the sand settles out to form sandbars and beaches. The settling of **sediment** is called **deposition**. Deposits of sand form beaches all over the world.

How did all this sand get on the beach?

19

Activate Prior Knowledge Have students share their experiences observing sand on beaches. What colors did they observe? What was the texture? Can you determine where the sand you observed came from?

Erosion

The beach sand might start on high mountain cliffs. Sometimes big chunks of rock fall off the sides of mountains. Gravity pulls rocks downhill. Other times landslides move rocks and soil downhill.

Rainwater flowing over the ground moves the pieces of broken rock. Water transports rocks into creeks. Water flowing in creeks transports broken rocks downstream. This is erosion.

Big chunks of rock can break off cliffs high in the mountains.

Flowing water transports, or moves, rocks down mountain creeks.

20

Creeks flow into rivers. Rivers have strong currents. Rivers can carry many sizes of rocks. The rocks tumble together and rub on the riverbed. The rocks break into smaller and smaller pieces. The smallest pieces are pebbles, gravel, sand, and silt. The farther the rocks move in the river, the smaller they get. They also get smoother and rounder as they are carried along.

Strong river currents move rocks farther downstream.

Smooth, round pebbles from a river

21

Compare and Contrast Ask students to describe and explain the differences between the rocks in the two inset photos.

SESSION 3

Deposition

Powerful **flood** waters move rocks of all sizes, even large boulders. When the water flowing in a river slows down, the rocks are dropped as sediment. Large rocks are the first to settle to the bottom.

Wherever a river flows into a lake, a bay, or the ocean, the water slows down. Sand is deposited near the mouth of the river. The sand can form sandbars, deltas, and beaches. Farther out are deposits of silt and clay.

Make Connections Have students discuss how the stream-table models compare to what happens in the natural world.

Large and small sediment deposited along a river after a flood

Notice the deposits of sand and silt where this river enters the lake.

22

Other Kinds of Erosion and Deposition

Wind blows sand and smaller pieces of rock from one place to another. Sometimes the wind blows hard enough to carry a lot of sand and dust. Wind can erode valuable farmland.

When the wind dies down, sand and dust are deposited far from their starting places. This is how sand dunes form. Three places in the United States where large sand dunes formed are Death Valley in California, White Sand Dunes in New Mexico, and Great Sand Dunes in Colorado.

Strong winds move earth materials from one place to another.

Sand dunes in Death Valley National Park, California

23

Sand deposited on a beach around large weathered rocks.

Glaciers are frozen rivers. Glaciers flow slowly through canyons. Rocks can be frozen in glaciers high in mountain canyons. The frozen rocks scrape the floor and sides of the canyon. Glaciers weather and erode V-shaped canyons into U-shaped valleys.

About 20,000 years ago in the western United States, glaciers scraped down mountain valleys. They crushed and ground up rock beneath them. At the same time, glaciers covered much of the Midwest. These sheets of ice were over 1.5 kilometers (km) thick. They changed much of the landscape by eroding the surface and depositing the rock material in new places.

What happens when sand finally makes it to the ocean? Is that the end of the erosion and deposition story? Not quite. Waves erode beaches and deposit sand in different places all the time. As waves crash on the beach, sand continues to weather. Sand gets finer and finer. Sand blowing in the air erodes the rocks and cliffs along the ocean shore. Erosion and deposition never stop.

Apply Understanding Have students discuss what the author means by "erosion and deposition never stop." Encourage students to ask questions they might investigate or research about the effects of erosion and deposition nearby.

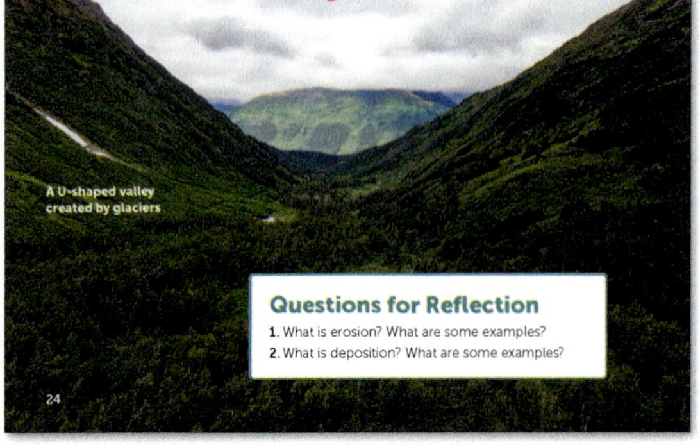

A U-shaped valley created by glaciers

Questions for Reflection

1. What is erosion? What are some examples?
2. What is deposition? What are some examples?

24

19 Discuss the reading

Review any unknown words or phrases or questions students have about the reading. Read aloud certain sections and use the discussion prompt and questions for reflection to check for student understanding. Encourage students to cite details and examples from the text to support their answers.

➜ *What is erosion? What are some examples?* [Moving of weathered eroded pieces. Rock pieces can be moved by flowing water in rivers or floods, wind, glaciers, and waves.]

➜ *What is deposition? What are some examples?* [The settling of sediment in a new location from its origin. Rock is deposited by slowing water, slowing wind, and melting glaciers.]

Explain,

What we observed in the model stream table can help us find out more about how earth materials move. Scientists often test ideas on a smaller size to help them think about what happens in a larger area of land over a long time. The smaller scale model can be used to think about the larger scale, like mountains and beaches.

Have students discuss this prompt with their group and then call on Reporters to share their groups' ideas. You might provide a white board or chart paper for groups to make a visual to present to the rest of the class.

➜ *Describe how rocks in the mountains become sand on a beach along the coast.* [Rock is weathered by roots, ice, abrasion, chemicals, and gravity. It is then eroded (transported) by streams and rivers until it reaches the sea. It is deposited at the shore when the river flows into the sea.]

ELA CONNECTION

These suggested strategies address these English Language Arts objectives.

Explain procedures, ideas, or concepts in a scientific text.

Add visual displays to presentations.

Crosscutting Concepts

- Scale, proportion, and quantity

PART 1 Erosion and Deposition

Reflecting on a Phenomenon

SESSION 3

20 Continue the explanation of the pile of earth materials phenomenon

Have students return to the section of their notebook where they are recording their thinking about the pile of earth material on the schoolyard. Ask them to add to their explanation of the phenomenon Show the visual in the Anchor Phenomenon Resource Video 1 as needed to facilitate students' thinking. Review the driving question:

➜ *What caused this pile of earth material to form?*

Ask,

➜ *How does the stream-table model help you to think about how the pile of earth material formed on the schoolyard?* [Water may have moved the earth materials from a higher place on the hill to the bottom just like the sand and clay moved when we poured water at the top of the stream table.]

21 View video as summary (optional)

The video *Erosion and Deposition* reviews the investigation and can be used before students share their notebook entries to facilitate the discussions.

22 Share notebook entries

Conclude Part 1 or start Part 2 by having students share notebook entries. Ask students to open their science notebooks to the most recent entry. Read the focus question together.

➜ *What causes weathered rock pieces to move from one place to another?*

Have students pair up with a student from another group to:

- share their stream-table drawings;
- discuss how rock materials get from mountains to a new location.

Encourage students to describe the cause-and-effect relationships involved in the processes.

Explain to students that their task is to help their partner improve their notebook entry. Have them take turns telling each other one thing they like about the entry and one thing they could do to make it better. Allow time for students to add to or revise their entries based on their partner's feedback.

ENGLISH LANGUAGE SUPPORT
See the Science-Centered Language Development chapter for strategies on sharing notebook entries.

Crosscutting Concepts
- Cause and effect

ELA CONNECTION
This suggested strategy addresses this English Language Arts objective.

Strengthen writing by revising.

INVESTIGATION 2

PART 2 Stream-Table Investigations

Science and Engineering Practices

- Developing and using models
- Planning and carrying out investigations
- Analyzing and interpreting data
- Constructing explanations
- Engaging in argument from evidence
- Obtaining, evaluating, and communicating information

Disciplinary Core Ideas

- ESS2.A: Earth materials and systems

Crosscutting Concepts

- Cause and effect
- Scale, proportion, and quantity
- Systems and system models
- Energy and matter

Aa

flood
floodplain
meander
mountain
slope

Teacher Background

Phenomenon Storyline

CONTINUED

LEARNING OBJECTIVES:

- **Students plan and carry out investigations using physical stream-table models to determine the effects of erosion and deposition on landforms.**
- **Students observe, record, and analyze data to determine cause-and-effect relationships about the rate and volume of erosion and the energy of moving water or wind.**

In Part 2, students continue to use stream-table model systems to explore variables that affect the outcomes of erosion and deposition such as the example they observed in the schoolyard phenomenon. The variables students investigate focus on slope and amount of water (flood). Here are some things students will confirm and find out to connect to the storyline.

- Weathered pieces of rock can be moved from place to place by water and wind (erosion). Water flowing downhill carries the rock pieces to new locations.
- Weathered rock particles settle out to form piles of earth materials (on a small scale) and to form new landforms (on a large scale). The settling out is called deposition.
- The steeper the slope of the hill or the more water that runs down the hill, the greater the erosion and deposition of earth material.

FOCUS QUESTION

How Does Slope Affect Erosion and Deposition?

Two variables affect the rate at which water erodes Earth's surface. Slope is the angle of the land over which the water flows. Because water is urged along by gravity, the steeper the angle of the surface, the faster the water flows. The energy of a moving object is a factor of both the mass of the object and its velocity. A given mass (volume) of water flowing across a gentle incline will wander across the land slowly, doing little to displace any of the particulate matter in its path. The same mass of water flowing over a steep section of land will find the lowest pathway and surge along that pathway at a much faster pace. The faster flow pushes larger particles of rock around and carries small particles along with the current. Therefore, the increased energy of the faster-moving water on a steep slope produces significantly more erosion than the same amount of water ambling slowly across a flatter landscape.

FOCUS QUESTION

How Do Floods Affect Erosion and Deposition?

Flood is the imprecise word used to describe a volume of water in a drainage system that exceeds the normal or typical volume in that system. Floods fill the streambed to overflowing. The mass of water during a flood is more than the environment can handle. During a flood, the energy of the mass of water in the stream channel may be several times greater than the energy of the water in the channel during times of normal flow.

Floods can propel boulders down stream channels. Boulders in motion gouge the streambed and smash smaller rocks to pieces. Flood energy amplifies both weathering and erosion. Tons of smaller sediment (sand and silt) are carried away. A tremendous amount of reshaping occurs, often not fully realized until the flood waters recede. Rivers can take a different course, new sandbars can be created, and vast amounts of mud (silt and clay) can be deposited in fields, in towns, and inside homes and other structures.

In Part 2, students manipulate the flow of water across the earth material plain, using the stream-table model. When they tilt the stream table to create a steeper slope, the water flows faster across the surface of the plain. By delivering the water in a more robust flow, students simulate a flood flowing over a plain. Students observe that a steeper slope that induces a faster flow of water increases erosion. Similarly, the increased volume of water induces greatly enhanced erosion. Students learn that fast-moving water has more energy than slow-moving water, and that a greater volume of water has more energy than a lesser volume of water. More energy produces more erosion.

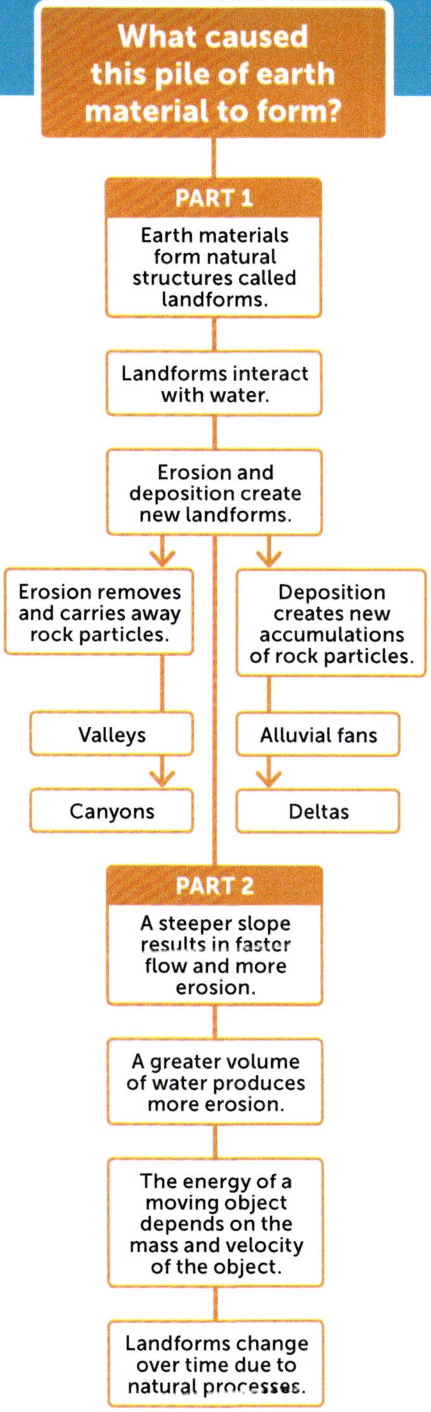

INVESTIGATION 2

PART 2 Stream-Table Investigations

Materials

PROVIDED EQUIPMENT

For each student

1 *FOSS Science Resources: Soils, Rocks, and Landforms*
"Landforms on Display"
"Where Do Rocks Come From?"
"Monumental Rocks" (optional)

For each group

1 Stream table with earth material
1 FOSS tray
1 Container, 1 L
1 Set of rock samples (sandstone, limestone, marble, basalt)
1 Basin
1 Ruler
1 Meter tape
1–2 Wood angles
1 Water source, standard (blue dot) or flood (red dot)

For the class

- Duct tape
- Cotton swabs
- Food coloring
- White, powdered clay

3 Plastic cups
1 Vial, 12 dr
- Pitcher
- Self-stick notes

SCHOOL-SUPPLIED ITEMS

For the class

- Water
- Safety goggles, gloves, and face masks for dust (optional)
- Paper towels and newspaper

1 Dustpan and broom

FOSSWEB RESOURCES

For each student

- Notebook sheet 9, *Stream-Table Observations* (2 copies)
- Notebook sheet 11, *Landform Definitions*
- Notebook sheet 12, *Stream-Table Investigation* (optional)
- Notebook sheet 13, *Response Sheet—Investigation 2*

For the class

- Word wall cards and equipment cards
- Teaching slides

For the teacher

- *Embedded Assessment Notes*

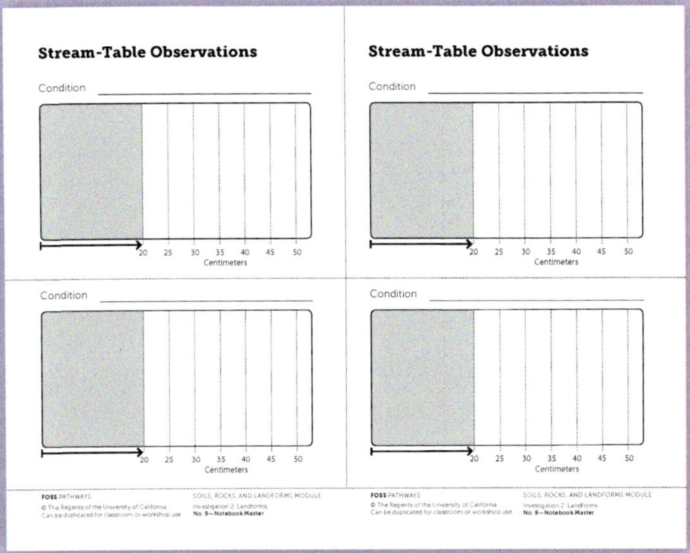

Stream-Table Observations

Condition ______

20 25 30 35 40 45 50
Centimeters

Condition ______

20 25 30 35 40 45 50
Centimeters

FOSS PATHWAYS
© The Regents of the University of California
Can be duplicated for classroom or workshop use
SOILS, ROCKS, AND LANDFORMS MODULE
Investigation 2: Landforms
No. 9—Notebook Master

Stream-Table Observations

Condition ______

20 25 30 35 40 45 50
Centimeters

Condition ______

20 25 30 35 40 45 50
Centimeters

FOSS PATHWAYS
© The Regents of the University of California
Can be duplicated for classroom or workshop use
SOILS, ROCKS, AND LANDFORMS MODULE
Investigation 2: Landforms
No. 9—Notebook Master

No. 9—Notebook Master

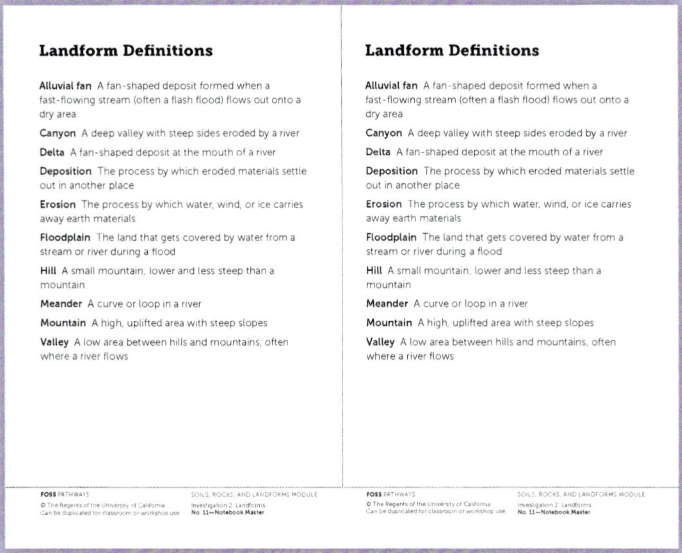

Landform Definitions

Alluvial fan A fan-shaped deposit formed when a fast-flowing stream (often a flash flood) flows out onto a dry area

Canyon A deep valley with steep sides eroded by a river

Delta A fan-shaped deposit at the mouth of a river

Deposition The process by which eroded materials settle out in another place

Erosion The process by which water, wind, or ice carries away earth materials

Floodplain The land that gets covered by water from a stream or river during a flood

Hill A small mountain; lower and less steep than a mountain

Meander A curve or loop in a river

Mountain A high, uplifted area with steep slopes

Valley A low area between hills and mountains, often where a river flows

FOSS PATHWAYS
© The Regents of the University of California
Can be duplicated for classroom or workshop use
SOILS, ROCKS, AND LANDFORMS MODULE
Investigation 2: Landforms
No. 11—Notebook Master

Landform Definitions

Alluvial fan A fan-shaped deposit formed when a fast-flowing stream (often a flash flood) flows out onto a dry area

Canyon A deep valley with steep sides eroded by a river

Delta A fan-shaped deposit at the mouth of a river

Deposition The process by which eroded materials settle out in another place

Erosion The process by which water, wind, or ice carries away earth materials

Floodplain The land that gets covered by water from a stream or river during a flood

Hill A small mountain; lower and less steep than a mountain

Meander A curve or loop in a river

Mountain A high, uplifted area with steep slopes

Valley A low area between hills and mountains, often where a river flows

FOSS PATHWAYS
© The Regents of the University of California
Can be duplicated for classroom or workshop use
SOILS, ROCKS, AND LANDFORMS MODULE
Investigation 2: Landforms
No. 11—Notebook Master

No. 11—Notebook Master

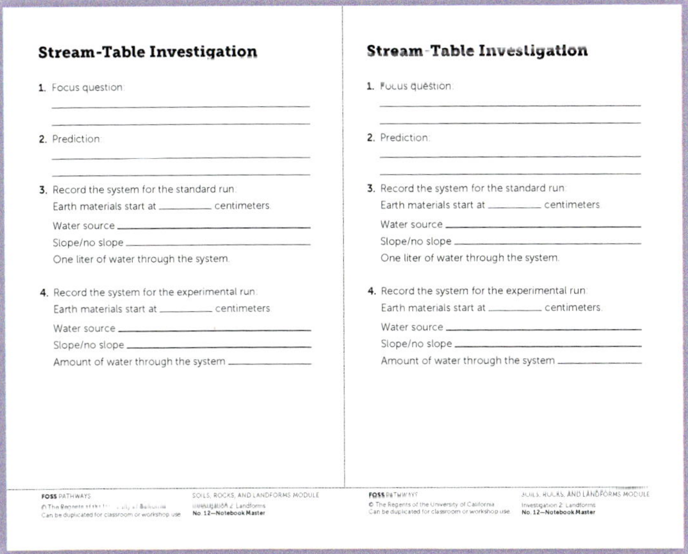

Stream-Table Investigation

1. Focus question: ______
2. Prediction: ______
3. Record the system for the standard run.
   Earth materials start at ______ centimeters.
   Water source ______
   Slope/no slope ______
   One liter of water through the system.
4. Record the system for the experimental run.
   Earth materials start at ______ centimeters.
   Water source ______
   Slope/no slope ______
   Amount of water through the system ______

FOSS PATHWAYS
© The Regents of the University of California
Can be duplicated for classroom or workshop use
SOILS, ROCKS, AND LANDFORMS MODULE
Investigation 2: Landforms
No. 12—Notebook Master

Stream-Table Investigation

1. Focus question: ______
2. Prediction: ______
3. Record the system for the standard run.
   Earth materials start at ______ centimeters.
   Water source ______
   Slope/no slope ______
   One liter of water through the system.
4. Record the system for the experimental run.
   Earth materials start at ______ centimeters.
   Water source ______
   Slope/no slope ______
   Amount of water through the system ______

FOSS PATHWAYS
© The Regents of the University of California
Can be duplicated for classroom or workshop use
SOILS, ROCKS, AND LANDFORMS MODULE
Investigation 2: Landforms
No. 12—Notebook Master

No. 12—Notebook Master

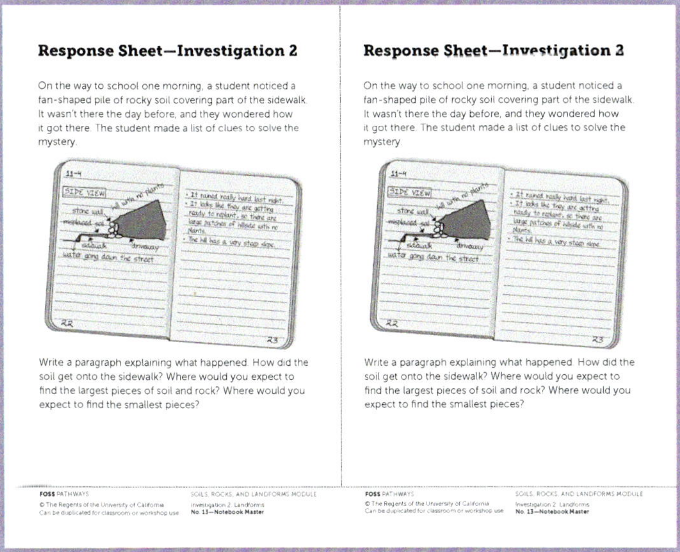

Response Sheet—Investigation 2

On the way to school one morning, a student noticed a fan-shaped pile of rocky soil covering part of the sidewalk. It wasn't there the day before, and they wondered how it got there. The student made a list of clues to solve the mystery.

Write a paragraph explaining what happened. How did the soil get onto the sidewalk? Where would you expect to find the largest pieces of soil and rock? Where would you expect to find the smallest pieces?

FOSS PATHWAYS
© The Regents of the University of California
Can be duplicated for classroom or workshop use
SOILS, ROCKS, AND LANDFORMS MODULE
Investigation 2: Landforms
No. 13—Notebook Master

Response Sheet—Investigation 2

On the way to school one morning, a student noticed a fan-shaped pile of rocky soil covering part of the sidewalk. It wasn't there the day before, and they wondered how it got there. The student made a list of clues to solve the mystery.

Write a paragraph explaining what happened. How did the soil get onto the sidewalk? Where would you expect to find the largest pieces of soil and rock? Where would you expect to find the smallest pieces?

FOSS PATHWAYS
© The Regents of the University of California
Can be duplicated for classroom or workshop use
SOILS, ROCKS, AND LANDFORMS MODULE
Investigation 2: Landforms
No. 13—Notebook Master

No. 13—Notebook Master

PART 2 Stream-Table Investigations

NOTE

Sessions 2 and 3 may run longer than a regular session while sessions 1 and 4 may be shorter than a regular session. If you do not have flexibility to extend your science time, here's what you might do.

Session 1, include Steps 3–4

Session 4, include Steps 17–19

Getting Ready

Schedule

4 Sessions

Session 1: Steps 1–2
Session 2: Steps 3–12 (Step 12 optional)
Session 3: Steps 13–19 (Step 19 optional)
Session 4: Steps 20–24 (Steps 22, 24 optional)

Review

- Background for Part 2
- FOSSweb Resources for this part
- Video: *Stream-Table Investigations* (optional)
- Tutorial: Soil Formation (optional)
- Readings: "Landforms on Display," "Where Do Rocks Come From?," and "Monument Rocks" (optional)
- Teacher Preparation Video

Print or Copy

For each student:

- Notebook sheet 9, *Stream-Table Observations*
- Notebook sheet 11, *Landform Definitions*
- Notebook sheet 12, *Stream-Table Investigation*
- Notebook sheet 13, *Response Sheet—Investigation 2*

For the teacher:

- *Embedded Assessment Notes*

Prepare Material

- Replenish earth material for stream table ____ A
- Prepare the notebook sheet introduction ____ B
- Prepare food coloring for stream table (optional) ____ C
- Prepare rock samples in trays ____ D

Plan for Assessment

- Review Step 18, "What to Look For" in the response sheet. Use *Response Sheet—Investigation 2* for a closer look at students' ideas about weathering, erosion, and deposition and how to use data to provide evidence for an explanation. The response to this prompt is also the explanation for the anchor phenomena of the pile of earth materials on the schoolyard. Plan to spend time having students reflect on their answers after you have reviewed them.

PREPARATION DETAILS

A Replenish earth materials for stream tables

After three or four uses of the earth materials in the stream tables, you may want to replenish the clay sediment. Work a 12-dram vial of white clay into the earth-material mixture of each stream table.

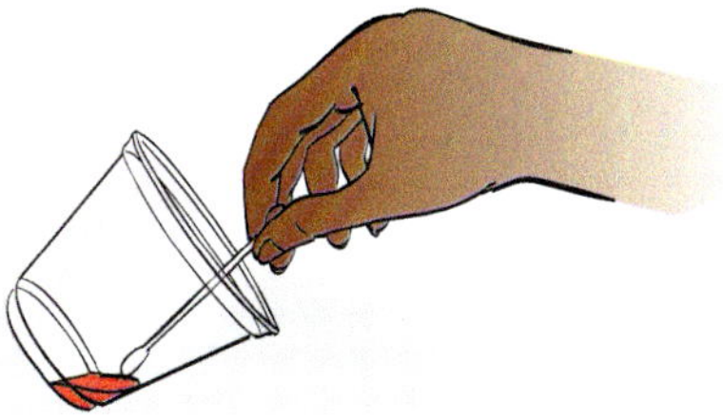

B Prepare the notebook sheet introduction

In this part, half the class will investigate slope condition and the other half will investigate flood condition. This approach is suggested so that both variables can be explored in one session. Prepare your introduction of the two different focus questions and the students use of notebook sheet 12, *Stream-Table Investigation*, to plan the two separate investigations. If students need more scaffolding to plan these investigations, consider conducting the investigations sequentially in two sessions, doing the slope conditions first and the flood investigation second.

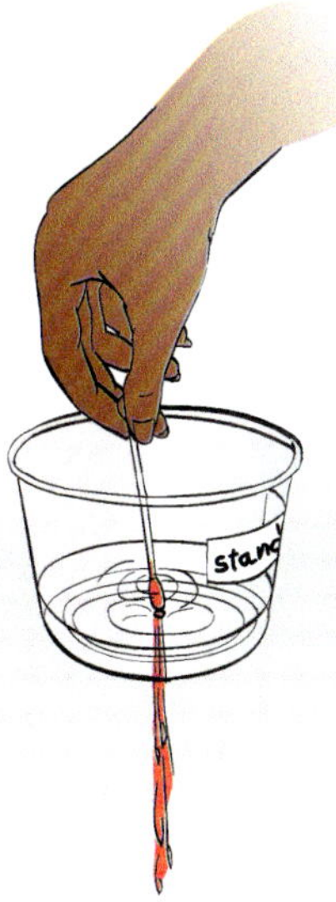

C Prepare food coloring for stream tables (optional)

Put 5 drops of red, blue, or green food coloring in cups to distribute during the stream-table investigations. Place one cotton swab in each cup. When you visit groups as they are running water through their stream tables, suggest that adding food coloring to the water will allow them to observe flow patterns more effectively. Do this by plunging a color-saturated swab into the water source for a few seconds, just above the hole in the water source. This will add an intense stream of color to the water flow.

D Prepare rock samples in trays

Each group needs a sample of four rocks for observations during the reading "Where Do Rocks Come From?" Prepare a tray of labeled rock samples for each group. Start by making a set of four labels (basalt, sandstone, limestone, marble) for each tray using self-stick notes. One rock label goes in each large depression of a tray. Place the corresponding rock sample on top of each label.

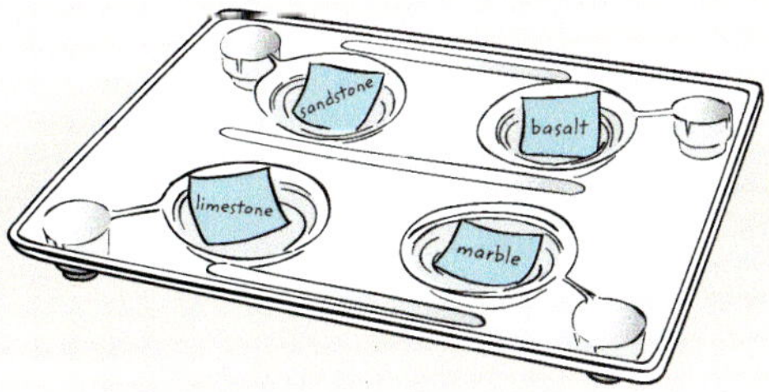

INVESTIGATION 2

PART 2 Stream-Table Investigations

Guiding the Investigation

SESSION 1 **45 MINUTES**

1 Revisit the pile of earth materials phenomenon

Revisit the phenomenon from Investigation 1 using the Anchor Phenomenon Resource Video 1. Ask students to discuss the processes of erosion and deposition and how it provides information to explain how the earth materials formed. Guide the discussion with these questions.

→ *What is erosion?* [The movement of earth material from one location to another.]

→ *What is deposition?* [The settling out of the earth material based on the size of the particle.]

→ *How does erosion and deposition contribute to the formation of landforms?* [The landform that is eroded is changed and new landforms might be formed over time.]

→ *What could we do with the stream tables to find out more about the pile of earth materials?* [Add more water, make the stream table a hill.]

Crosscutting Concepts

- Cause and effect
- Scale, proportion, and quantity

Review notebook sheet 10, *Landform Vocabulary*, as a class and make a concept grid on chart paper. Add the landforms in the first column and title the other columns as shown below. Tell students that they will find out more about how landforms are formed and add that information to the concept grid. They can also speculate about how human actions might impact the landforms.

Remind students about the scale differences between the stream-table model, the pile of earth material at the bottom of the hill in the schoolyard scenario, and the landform of a canyon of a valley. The processes that cause change in these systems may be very similar, but the size and time frame are at different scales. Scientists often test ideas on a smaller size to help them think about what happens in a larger area of land or over a long time.

Landform	Characteristics	How it was formed	Human impact
Alluvial fan			
Canyon			
Delta			
Floodplain			
Meander			
Valley			

PART 2 Stream-Table Investigations

Interpreting Information

SESSION 1

Science and Engineering Practices

- Analyzing and interpreting data
- Obtaining, evaluating, and communicating information

ELA CONNECTION

This suggested strategy addresses this English Language Arts objective.

Interpret information presented visually.

NOTE

Students will revisit and discuss the last four pages of this article in Investigation 4, Part 1, where they investigate earthquakes and volcanoes.

2 Read "Landforms on Display"

Have students open their *FOSS Science Resources* books to "Landforms on Display" and focus on the photos of landforms caused by weathering, erosion, and deposition.

Give students a moment to look through the photographs and discuss any personal connections with a reading partner. Refer to the concept grid and have students search for these particular landforms in the article. Together as a class, fill in the characteristics and information about how the landform or land feature was formed.

As a class, review the other landforms and discuss questions students have. Use the questions and prompts as need to guide the discussion.

Visualize Have students visualize how this arch formed over time. How is the process the same and different from arches seen near the ocean?

Landforms on Display

Landforms Created by Weathering and Erosion

ARCH A curved rock that forms when chemical and/or physical weathering weakens the center, and the rock erodes. Arches can form on the land or near the coast where waves batter and erode the center of the rock.

25

Landforms Created by Weathering and Erosion

MESA A single, wide, flat-topped hill having at least one steep side.

BUTTE A hill with steep sides and a small, flat top. A butte is smaller than a mesa.

HOODOO Statue-like shapes resulting from differential weathering and erosion.

CANYON A V-shaped gorge with steep sides eroded by a stream.

GORGE A narrow, steep-sided canyon.

Many landforms in the American Southwest are the result of weathering and erosion.

26

27

Apply Understanding Discuss what these landforms may have looked like a long time ago and what they might look like way into the future. What might change? What causes these landforms to change?

PART 2 Stream-Table Investigations

SESSION 1

Make Connections Have students compare these landforms to the ones made in the stream-table models. How might they use the stream tables to investigate what causes these landforms?

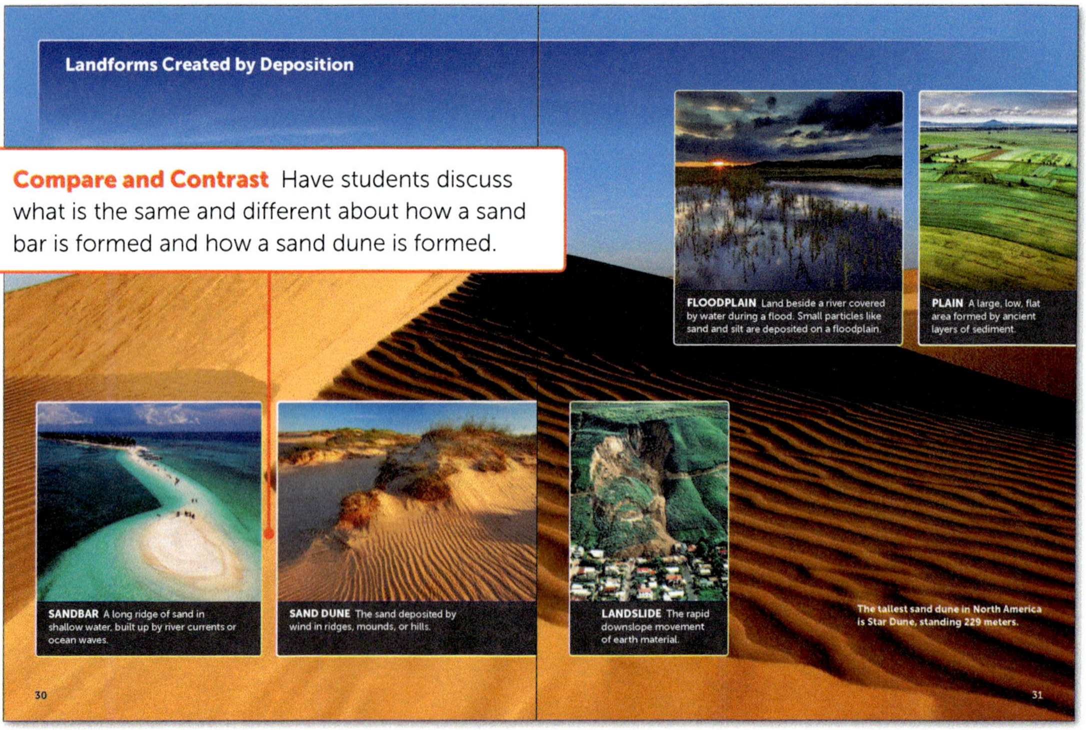

Compare and Contrast Have students discuss what is the same and different about how a sand bar is formed and how a sand dune is formed.

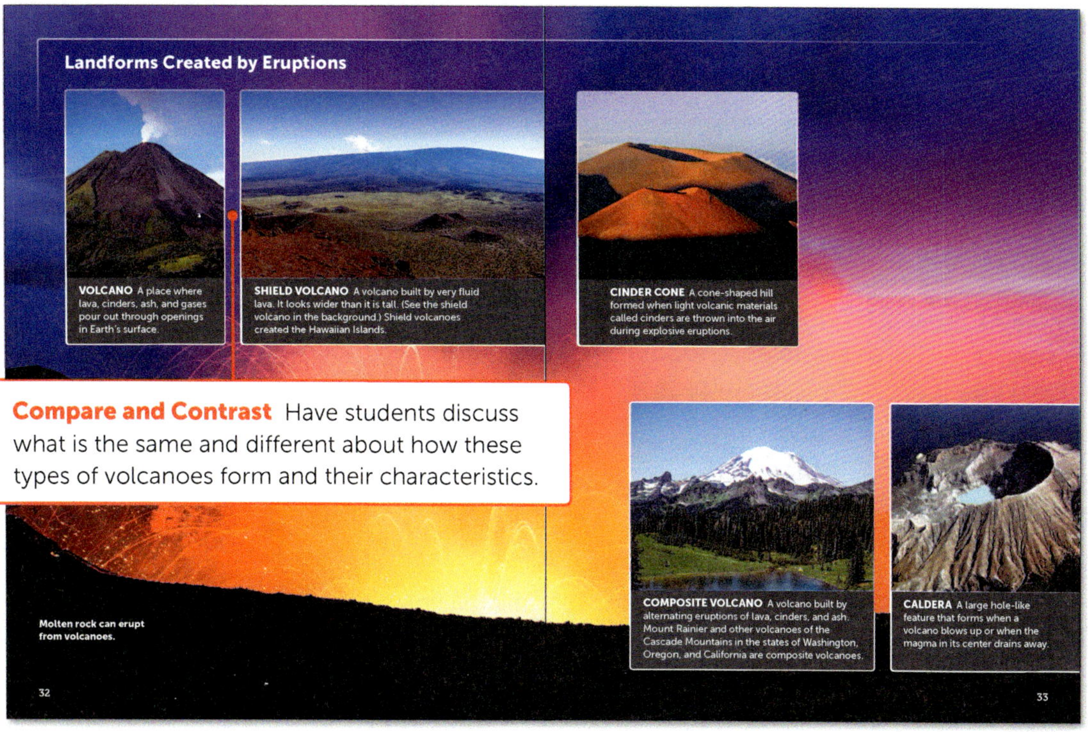

Landforms Created by Eruptions

VOLCANO A place where lava, cinders, ash, and gases pour out through openings in Earth's surface.

SHIELD VOLCANO A volcano built by very fluid lava. It looks wider than it is tall. (See the shield volcano in the background.) Shield volcanoes created the Hawaiian Islands.

CINDER CONE A cone-shaped hill formed when light volcanic materials called cinders are thrown into the air during explosive eruptions.

COMPOSITE VOLCANO A volcano built by alternating eruptions of lava, cinders, and ash. Mount Rainier and other volcanoes of the Cascade Mountains in the states of Washington, Oregon, and California are composite volcanoes.

CALDERA A large hole-like feature that forms when a volcano blows up or when the magma in its center drains away.

Molten rock can erupt from volcanoes.

32

33

Compare and Contrast Have students discuss what is the same and different about how these types of volcanoes form and their characteristics.

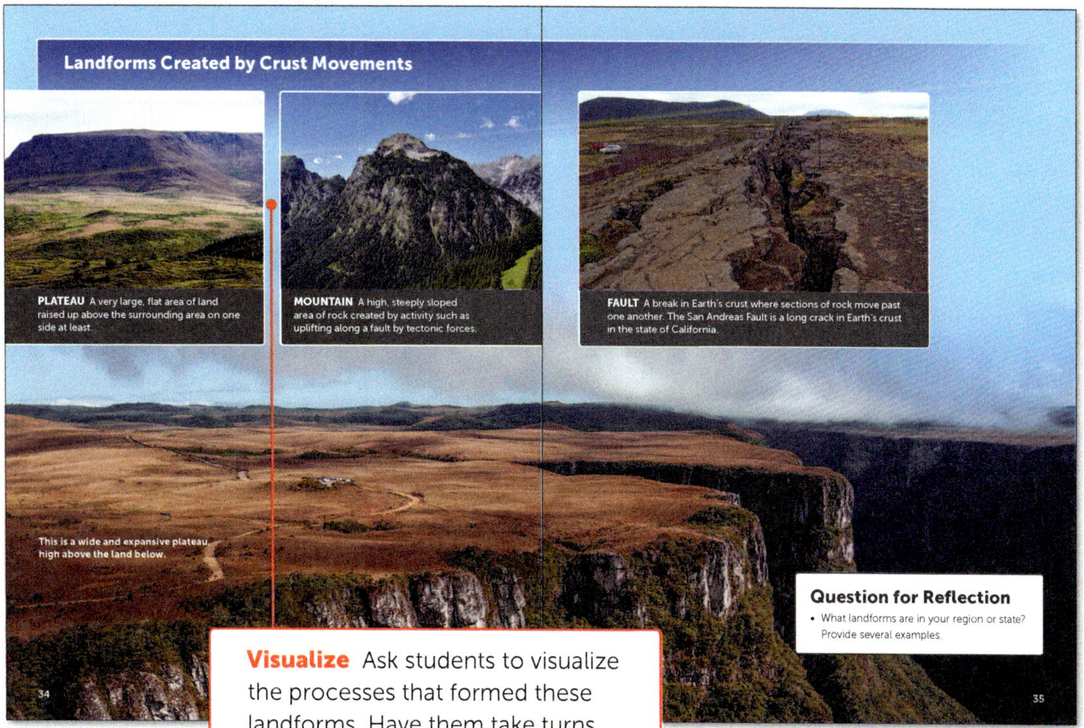

Landforms Created by Crust Movements

PLATEAU A very large, flat area of land raised up above the surrounding area on one side at least.

MOUNTAIN A high, steeply sloped area of rock created by activity such as uplifting along a fault by tectonic forces.

FAULT A break in Earth's crust where sections of rock move past one another. The San Andreas Fault is a long crack in Earth's crust in the state of California.

This is a wide and expansive plateau high above the land below.

Question for Reflection

- What landforms are in your region or state? Provide several examples.

34

35

Visualize Ask students to visualize the processes that formed these landforms. Have them take turns acting out how tectonic forces might have created these landforms.

PART 2 Stream-Table Investigations

Guiding the Investigation

NOTE

This part is written for you to guide the investigation planning for both focus questions, given that most elementary students have little experience. Keep these guidelines in mind, but let students lead the investigation-planning discussion. If your students have experience planning investigations, let them go ahead on their own, but check plans and setup before they start the water flowing. To meet the needs of students, you can consider breaking Session 2 into two sequential sessions as described in Getting Ready.

SESSION 2 **45 MINUTES**

3 Introduce the focus questions

FOCUS QUESTION

How does slope affect erosion and deposition? How do floods affect erosion and deposition?

Tell students that today they get to test two of their ideas: **slope** of the land and **flood**. Project or write both focus questions on the board as you read them aloud. Students can wait to record them.

➜ *How does slope affect erosion and deposition?*

➜ *How do floods affect erosion and deposition?*

Explain that half the class will focus on slope and the other half will focus on floods. The class will share results so that everyone will be able to answer both focus questions.

Two stream tables are needed, one for a control setup and one for the experimental condition. Pair up groups so they can run two stream tables side by side. One group will run a standard stream table (as they did in Part 1) and the other group will setup a condition one to test. By working in group pairs, students will be able to easily compare the standard result to the condition under investigation—slope or flood.

4 Plan the slope investigation (first focus question)

Introduce the slope condition and plan the slope investigations as a class. Project notebook sheet 12, *Stream-Table Investigation*, and demonstrate how to use it to record the plan. These points should be included.

- The focus question: How does slope affect erosion and deposition?
- Prediction: How do you think increasing the slope will affect erosion and deposition? (Use a sentence frame if needed: *I think* _____ *because* _____.)
- System variables to control: The stream-table models are systems and have parts that function together. All stream-table models will have earth materials pushed to 20 cm, and will run exactly 1 L of water through the system, and the water flow in both stream tables will start at the same time.
- System variable to change for slope investigation: One group (standard stream table) will leave their stream table flat on the table with no slope (as they did in Part 1 with a pencil to hold up the water-source end). The other group (slope stream table) will prop up the water-source end of the stream table using two wood angles.

Science and Engineering Practices

- Developing and using models
- Planning and carrying out investigations
- Constructing explanations

Crosscutting Concepts

- Systems and system models

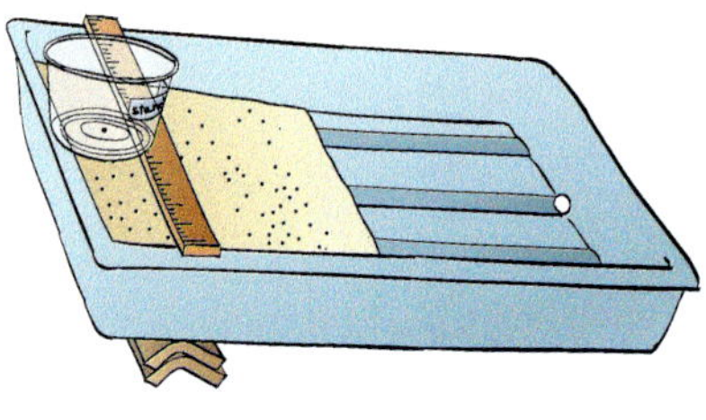

Note two wood angles are used by the "slope" groups.

5 Plan the flood investigation (second focus question)

Introduce the flood condition. Ask students if they know what a flood is. If they need help, here are some key points to help guide the discussion.

a. Floods are caused by excessive runoff from rain or snowmelt.

b. If soil is frozen or saturated with water, water can't soak in, so runoff is greater.

c. Because there is so much water in a flood, land that is normally dry gets covered with water.

Plan the flood investigations as a class. Hold up the flood water source and show them the red dot and the larger hole on the bottom. These points should be included in the plan.

- The focus question: How do floods affect erosion and deposition?
- Prediction: How do you think increasing the amount of water (using a water source with a bigger hole on the bottom) will affect erosion and deposition? (Use a sentence frame if needed: *I think* ______ *because* ______.)
- System variables to control: The stream-table models are systems and have parts that function together. All stream-table models will have earth materials pushed to 20 cm, and will run exactly 1 L of water through the system, and the water flow in both stream tables will start at the same time.
- System variable to change for flood investigation: One group (standard stream table) will leave their stream table flat on the table and use the standard water source (as they did in Part 1 with a pencil to hold up the water-source end). The other group (flood stream table) will use the flood water-source (red dot) and also only the pencil to prop up the stream-table end.

6 Record stream-table systems plan

Assign groups to investigate either slope or flood. Distribute copies of notebook sheet 12, *Stream-Table Investigation*, and have students complete the sheet before they set up their systems. Have students work together in their groups to complete the sheet.

Have copies of notebook sheet 9, *Stream-Table Observations*, available for students to pick up and record results of the standard and either slope or flood stream-table model systems.

INVESTIGATION 2

PART 2 Stream-Table Investigations

SESSION 2

7 Set up stream-table model systems

Have Getters get the materials they will need to run the stream tables. Give students a few minutes to get the earth material ready. They might need to break up the material if it has solidified. Adding a little water will speed up the process.

When the groups have their earth material ready, all the parts of their model systems assembled, and newspaper strategically deployed, make sure that the slope groups have placed two wood angles under their stream tables to prop up the water-source end of the tray.

8 Carry out the slope and flood investigations

Send Getters back to the materials station for 1 L of water and let the trials begin. Remind students that both groups should start the water running at the same time for the stream-tables model systems that are running side by side. When all of the water has flowed through the table, they should begin to draw and write their observations.

9 Guide the recording of the stream-table runs

As you visit the groups, have students compare the slope or flood results to the standard stream-table results. Ask students a few questions to help guide their recording and to continue to build vocabulary and construct explanations.

→ *Where did erosion occur? Did the slope (or flood) make a difference in how much earth material eroded?* [Erosion occurred in the canyon; more material eroded in the steeper tray.]

→ *Did you notice any difference in the time it took for erosion or deposition?* [Things happened faster in the slope tray.]

→ *Did the slope or flood cause a difference in the shape of the canyon that was left after the earth materials were carried away?* [The canyon is usually deeper and has steeper sides.]

→ *How far did the eroded materials travel in the slope (or flood) conditions compared to the standard stream table conditions?* [The earth material traveled farther and/or faster.]

Have students complete drawings and descriptions of the two stream tables.

MATERIALS FOR STEP 7

- Stream tables, with earth materials
- Water sources, standard
- Water sources, flood
- Rulers
- Wood angles
- Containers, 1 L
- Basins
- Meter tapes
- Newspapers
- Duct tape
- Water

Science and Engineering Practices

- Planning and carrying out investigations
- Analyzing and interpreting data
- Constructing explanations

NOTE

This important idea may not be obvious to students. You may want to ask questions to help students think about a ball rolling on a gentle slope compared to one rolling on a steep slope. Any object moving in response to gravity down a steeper slope will move faster and have more energy than the same object moving down a gentler slope. Faster = more energy.

Crosscutting Concepts

- Cause and effect

10 Visit other stream-table investigations

After all the investigations are complete, have the students working on slope observe the results of the flood investigations and vice versa. Provide additional copies of notebook sheet 9, *Stream-Table Observations*, for students to use for recording the other investigations. Decide on a rotation method for making these observations for the other system.

Science and Engineering Practices

- Analyzing and interpreting data

11 Finish recordings and take photos

At the end of this session, each student should have a record of the results of the slope investigation and the flood investigation in their notebooks with a comparison to the standard systems. They will use these notes for a sense-making discussion in the next session.

Take photos of representative stream-table model systems to project and share during the next session.

12 Run more water through each system (optional)

If there is time, allows students to run a second liter of water through each stream-table model system. Provide the food coloring for students to use. There is no need to reset the earth material for this experience, just let the students observe the results. Students can record observations in their notebooks. Take photos of representative stream-table model systems.

Have students place a small piece of duct tape over the drain hole in their trays and stack the trays crisscross in an appropriate location. Crisscross the stream tables for storage.

Have students return everything but the stream table with the earth material to the materials station.

Allow the basins of water to settle after running the stream tables. Ask the Starters to recycle the water in the basins outdoors and clean the tabletops. Stir the clay back into the stream-table earth material.

Gritty materials should be rinsed in a basin of water (never down the sink) and set out to dry.

INVESTIGATION 2

PART 2 Stream-Table Investigations

flood
floodplain
meander
mountain
slope

Science and Engineering Practices

- Analyzing and interpreting data
- Constructing explanations

Crosscutting Concepts

- Cause and effect
- Energy and matter

SESSION 3 **45 MINUTES**

13 Review vocabulary

Review the new vocabulary introduced in this part. Have students turn to notebook sheet 10, *Landform Vocabulary*. Have them glue notebook sheet 11, *Landform Definitions*, on the opposite page. In their groups, have students review the landform definitions and place new landform terms in the appropriate place on the graphic.

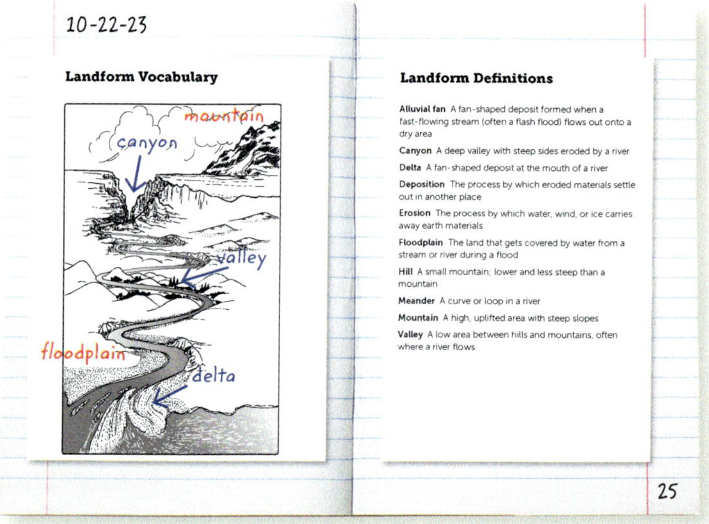

14 Have a sense-making discussion

Have students look at their drawings of the standard stream table. Alternatively, if students captured images, have those available for comparison. Ask students to construct explanations to these questions.

→ *How does the erosion of the stream table with a slope or flood compare to the standard stream table?* [For the steeper slope, the earth material eroded faster, moved farther, and formed a deeper valley—a canyon. Flood conditions also result in more erosion; broader, deeper, straighter canyons, and more extensive deposition that reached farther into the basin.]

→ *What causes more erosion to happen in the flood than in the standard?* [The water comes down faster.]

→ *When water comes down faster during a flood, the water has more energy than a normal rainfall. What effect does this increased energy have on the earth materials?* [The more energy, the more erosion.]

→ *Does the water on the slope have more energy than the water on the flat stream table? What data do you have to support that idea?* [More energy; there was more erosion (deeper canyons, earth materials moved further).]

→ *Why do you think there was more erosion in the steeper tray?* [When the land was steeper, the water flowed faster. Faster flowing water has more energy to move sand and clay farther along the stream or river channel.]

→ *How does this help you explain what happened in the photos in the "Landforms on Display" article?*

Students should find that in the stream-table model system set at a steeper slope, the earth material eroded faster, moved farther, and formed a deeper valley—a canyon. Flood conditions also resulted in more erosion; broader, deeper, straighter canyons, and more extensive deposition that reached farther into the basin. The flood eroded more earth material than the standard and sloped stream tables because a larger mass of moving water has more energy than a smaller mass of moving water.

15 Respond to the two focus questions

Have students write responses to the focus questions in their notebooks.

→ *How does slope affect erosion and deposition?*

→ *How do floods affect erosion and deposition?*

16 Connect to the anchor phenomenon

Ask students,

→ *How might erosion and deposition contribute to the kinds and sizes of rock material that you find in soils?* [The kinds of rock in soil depends on the kind of parent rock (source rock) and how the rock has weathered and moved. As rocks erode, they tumble against one another; in a river or flood, more pieces break off and get smaller. Particles are deposited as part of the soil.]

→ *How does this help us explain how the pile of earth material at the bottom of the hill got on the schoolyard?*

Tell students,

Normal weathering and erosion causes changes to Earth's surface a tiny bit at a time. It can take thousands of years to create a valley or alluvial fan.

A flood is an event that causes changes to Earth's surface in a short period of time. Floods can destroy existing landforms and create new landforms in hours or days. A flood can erode and redeposit more rock material in 10 hours than normal water flow can erode and deposit in 10 years. Gravity pulls water and eroded earth materials on mountains and other sloped surfaces downhill where the materials are deposited.

ENGLISH LANGUAGE SUPPORT

For students who need scaffolding, provide a sentence frame such as:

Slope causes _____. When we compared _____ to _____, we observed _____. This happens because _____.

Floods cause _____. When we compared _____ to _____, we observed _____.

NOTE

Make sure students understand that soil formation takes many years, in fact so many that some people consider soil a nonrenewable resource.

Crosscutting Concepts

- Cause and effect

SESSION 3

17 Identify soils from Investigation 1

NOTE

Bring out the four soil vials you saved from Investigation 1 and show them to students to remind them of what they looked like.

Have students turn back to the entries in their science notebooks from Investigation 1 where they first attempted to identify where each soil studied came from: desert, delta, forest, or mountaintop.

Have them discuss these soils in their groups, providing additional evidence for their arguments from what they have learned about erosion and deposition. Have them share their final decisions with the class, then reveal the soil sources (locations).

Science and Engineering Practices

- Engaging in argument from evidence

Soil 1—River delta (Evidence: the smallest earth materials, carried a long way and then deposited in a delta)

Soil 2—Mountaintop (Evidence: large pieces of earth material, not a lot of humus)

Soil 3—Desert (Evidence: mostly sand and gravel, not a lot of humus)

Soil 4—Forest (Evidence: lots of humus and different sizes of earth materials)

Ask students to drain and stack the stream tables. Crisscross the stream tables for storage. Gritty materials should be rinsed in a basin of water (never down the sink) and set out to dry.

18 Assess progress: response sheet

Science and Engineering Practices

- Analyzing and interpreting data
- Constructing explanations

Disciplinary Core Ideas

- ESS2.A: Earth materials and systems

Crosscutting Concepts

- Cause and effect

Distribute a copy of notebook sheet 13, *Response Sheet—Investigation 2* to each student. This sheet is an opportunity for students to practice their explanation about the anchor phenomenon 1 scenario. The assessment should be completed during class. Collect the response sheets in student notebooks after class and check students' ability to analyze and interpret the information on the sheet and construct an explanation based on data.

WHAT TO LOOK FOR

- *Students explain that soil on the sidewalk is a result of erosion caused by rainwater flowing on the hillside and deposition on the sidewalk.*
- *Students explain that based on the analysis of the stream-table system data, larger pieces of soil will be deposited near the rock wall, and smaller pieces will travel farther along the water flow.*

19 View video *Stream-Table Investigations* (optional)

Have students view the video to gain more experience with observing and recording the effects of changing variables of slope of the land and amount of river flow (flood conditions).

Interpreting Information

SESSION 4 **45 MINUTES**

20 Read "Where Do Rocks Come From?"

Have students read "Where Do Rocks Come From?" to obtain and communicate information about the parent rocks that are broken into pieces to become part of soil. In this article, students learn about the three types of rock—igneous, sedimentary, and metamorphic rocks.

Before reading, have students preview the text by looking at and discussing the diagrams and photos with a partner. Students might jigsaw the article. Have one person in each group read one section and then present the main ideas to the rest of the group. (See the Science-Centered Language Development chapter for more information about this reading strategy.)

Distribute a tray of rock samples to each group. Encourage students to observe each rock sample when it appears in the text or photos. They should also review the rocks when they study the table of photos on page 41.

Science and Engineering Practices

- Obtaining, evaluating, and communicating information

ELA CONNECTION

These suggested strategies address these English Language Arts objectives.

Determine the main idea of a text and explain how it is supported by key details; summarize text.

Explain events, procedures, ideas, or concepts in a scientific text, including what happened and why, based on specific information in the text.

MATERIALS FOR STEP 20

- Trays, each with four labeled rock samples (basalt, sandstone, limestone, marble)

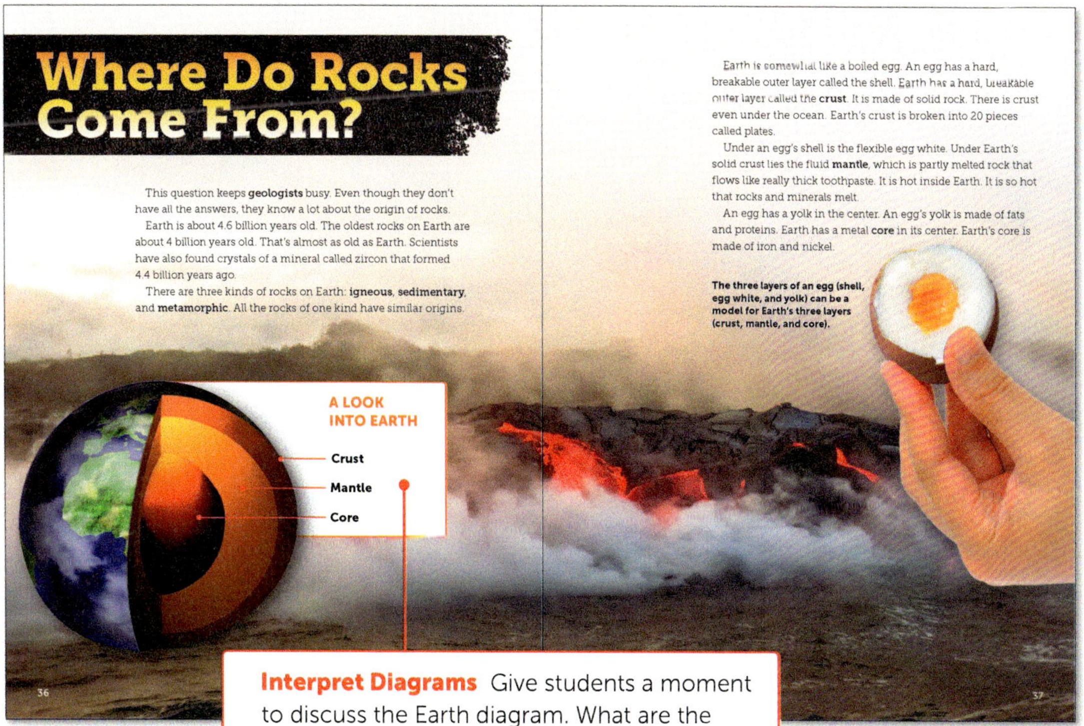

Where Do Rocks Come From?

This question keeps **geologists** busy. Even though they don't have all the answers, they know a lot about the origin of rocks.

Earth is about 4.6 billion years old. The oldest rocks on Earth are about 4 billion years old. That's almost as old as Earth. Scientists have also found crystals of a mineral called zircon that formed 4.4 billion years ago.

There are three kinds of rocks on Earth: **igneous**, **sedimentary**, and **metamorphic**. All the rocks of one kind have similar origins.

36

Earth is somewhat like a boiled egg. An egg has a hard, breakable outer layer called the shell. Earth has a hard, breakable outer layer called the **crust**. It is made of solid rock. There is crust even under the ocean. Earth's crust is broken into 20 pieces called plates.

Under an egg's shell is the flexible egg white. Under Earth's solid crust lies the fluid **mantle**, which is partly melted rock that flows like really thick toothpaste. It is hot inside Earth. It is so hot that rocks and minerals melt.

An egg has a yolk in the center. An egg's yolk is made of fats and proteins. Earth has a metal **core** in its center. Earth's core is made of iron and nickel.

The three layers of an egg (shell, egg white, and yolk) can be a model for Earth's three layers (crust, mantle, and core).

37

Interpret Diagrams Give students a moment to discuss the Earth diagram. What are the characteristics of each layer? How might these layers interact?

PART 2 Stream-Table Investigations

SESSION 4

Igneous Rocks

Igneous rocks start out as melted rock deep in Earth's crust. Sometimes the melted rock, called **magma**, comes to the surface in **volcanoes**. When the magma pours out on the surface, it is called **lava**. When lava cools and hardens, it forms new rocks. The basalt you tested for calcite is volcanic igneous rock. Much of the rock in the Cascade Mountains of the Pacific Northwest is basalt.

In other places, magma cooled slowly and hardened below the surface. Earthquakes and other tectonic events in Earth's crust brought these igneous rocks to the surface thousands of years later. The **granite** you studied cooled below Earth's surface. The Sierra Nevada Mountains in central California and the Rocky Mountains in Colorado, Wyoming, and Montana are mostly granite mountains.

Basalt

Granite

The lava from volcanoes forms igneous rocks.

38

Sedimentary Rocks

Sedimentary rocks form from bits and pieces of rocks and minerals. **Sandstone** is an example of a sedimentary rock. Sandstone starts as big rocks in the mountains. Over time, the rocks crack and break into smaller pieces in a process called weathering.

Loose rocks roll downhill and break into smaller pieces. This process can take millions of years. Pieces might end up in streams and rivers. The pieces tumble around and break into smaller and smaller pieces. Eventually the rocks from the mountain are reduced to tiny pieces of sand.

Sandstone

Sandstone in places like the Valley of Fire in Nevada show beautiful bands of color.

39

Summarize Have students take turns paraphrasing how igneous rocks are formed with a partner.

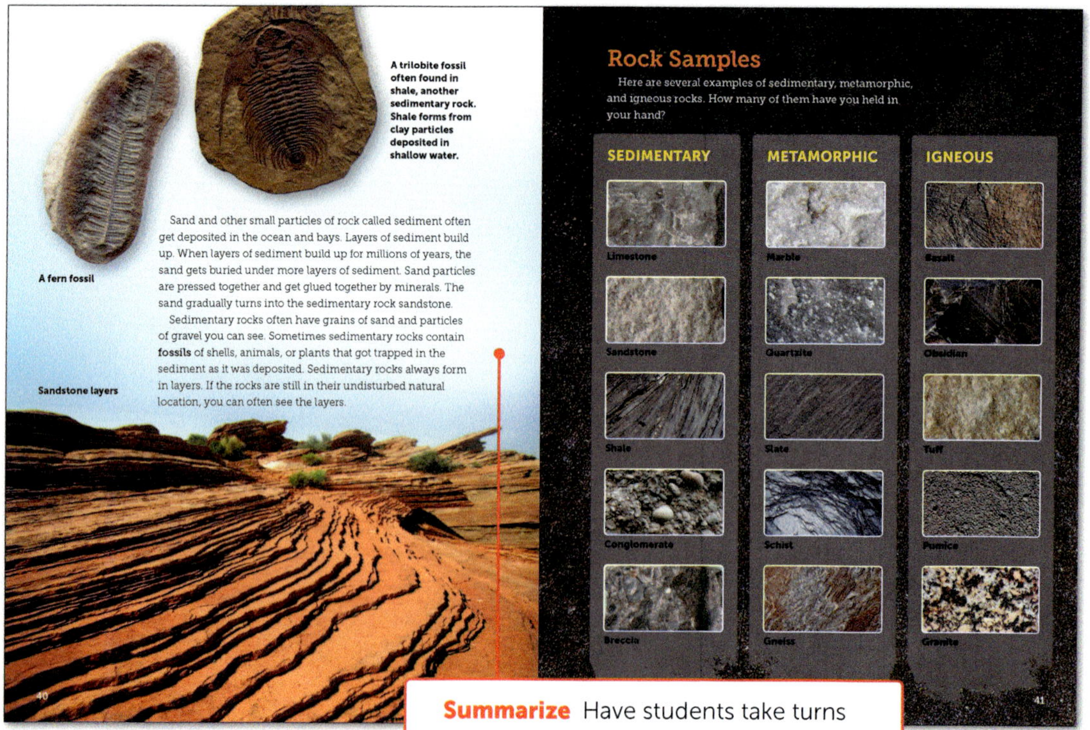

A trilobite fossil often found in shale, another sedimentary rock. Shale forms from clay particles deposited in shallow water.

A fern fossil

Sand and other small particles of rock called sediment often get deposited in the ocean and bays. Layers of sediment build up. When layers of sediment build up for millions of years, the sand gets buried under more layers of sediment. Sand particles are pressed together and get glued together by minerals. The sand gradually turns into the sedimentary rock sandstone.

Sedimentary rocks often have grains of sand and particles of gravel you can see. Sometimes sedimentary rocks contain **fossils** of shells, animals, or plants that got trapped in the sediment as it was deposited. Sedimentary rocks always form in layers. If the rocks are still in their undisturbed natural location, you can often see the layers.

Sandstone layers

40

Rock Samples

Here are several examples of sedimentary, metamorphic, and igneous rocks. How many of them have you held in your hand?

SEDIMENTARY	METAMORPHIC	IGNEOUS
Limestone	Marble	Basalt
Sandstone	Quartzite	Obsidian
Shale	Slate	Tuff
Conglomerate	Schist	Pumice
Breccia	Gneiss	Granite

41

Summarize Have students take turns paraphrasing how sedimentary rocks are formed with a partner.

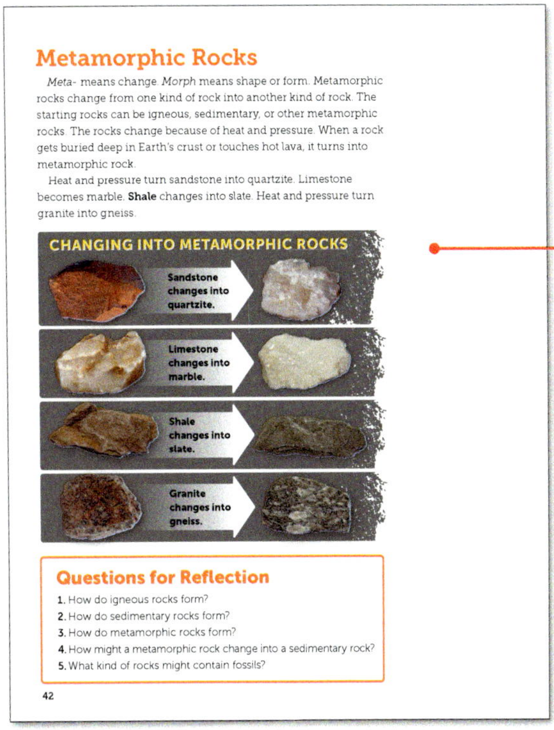

Metamorphic Rocks

Meta- means change. *Morph* means shape or form. Metamorphic rocks change from one kind of rock into another kind of rock. The starting rocks can be igneous, sedimentary, or other metamorphic rocks. The rocks change because of heat and pressure. When a rock gets buried deep in Earth's crust or touches hot lava, it turns into metamorphic rock.

Heat and pressure turn sandstone into quartzite. Limestone becomes marble. **Shale** changes into slate. Heat and pressure turn granite into gneiss.

Questions for Reflection

1. How do igneous rocks form?
2. How do sedimentary rocks form?
3. How do metamorphic rocks form?
4. How might a metamorphic rock change into a sedimentary rock?
5. What kind of rocks might contain fossils?

42

Summarize Have students take turns paraphrasing how metamorphic rocks are formed with a partner.

21 Discuss the reading

Have groups discuss and share what they learned from the article. If needed, ask,

- *How do igneous rocks form?* [They start out as melted rock in Earth's mantle. They can come to the surface in volcanoes and pour out as lava, or they can cool beneath the surface of Earth.]
- *How do sedimentary rocks form?* [They form from bits and pieces of rocks and minerals. The sediments form layers that are pressed and cemented together over millions of years.]
- *How do metamorphic rocks form?* [These rocks form when one kind of rock changes into another kind of rock due to intense heat and pressure deep in Earth's crust or under lava.]
- *How might a metamorphic rock change into a sedimentary rock?* [If a metamorphic rock weathered, chemically or physically, it could break apart and form sediment that could become a sedimentary rock.]
- *What kind of rocks might contain fossils?* [Sedimentary rocks.]

ELA CONNECTION

This suggested strategy addresses this English Language Arts objective.

Explain events, procedures, ideas, or concepts in a scientific text, including what happened and why, based on specific information in the text.

PART 2 Stream-Table Investigations

ELA CONNECTION

These suggested strategies address these English Language Arts objectives.

Determine the main idea of a text and explain how it is supported by key details; summarize the text.

Gather relevant information from print; take notes and categorize information.

Science and Engineering Practices

- Obtaining, evaluating, and communicating information

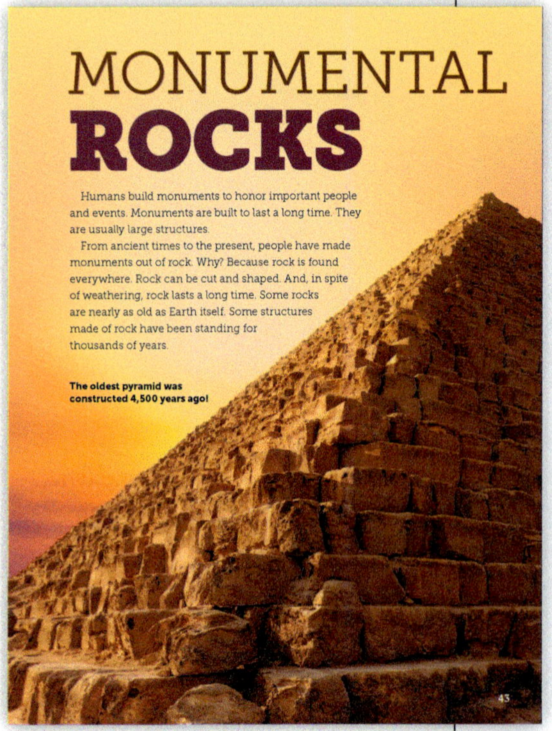

MONUMENTAL ROCKS

Humans build monuments to honor important people and events. Monuments are built to last a long time. They are usually large structures.

From ancient times to the present, people have made monuments out of rock. Why? Because rock is found everywhere. Rock can be cut and shaped. And, in spite of weathering, rock lasts a long time. Some rocks are nearly as old as Earth itself. Some structures made of rock have been standing for thousands of years.

The oldest pyramid was constructed 4,500 years ago!

43

SESSION 4

22 Read "Monumental Rocks" (optional)

Have students read the article "Monumental Rocks" to obtain, evaluate, and communicate information after completing the active investigation.

Before the reading, tell students that the "Monumental Rocks" article describes four monuments of worldwide significance. Read the introductory paragraph aloud and ask students to think about what is the same about these monuments and what is different. Continue reading aloud or have students read independently. Pause after each section to have students take turns summarizing the text.

Make a class graphic organizer, or have students work as a group or in pairs to compare and contrast the four monuments.

Compare and Contrast Four Monuments

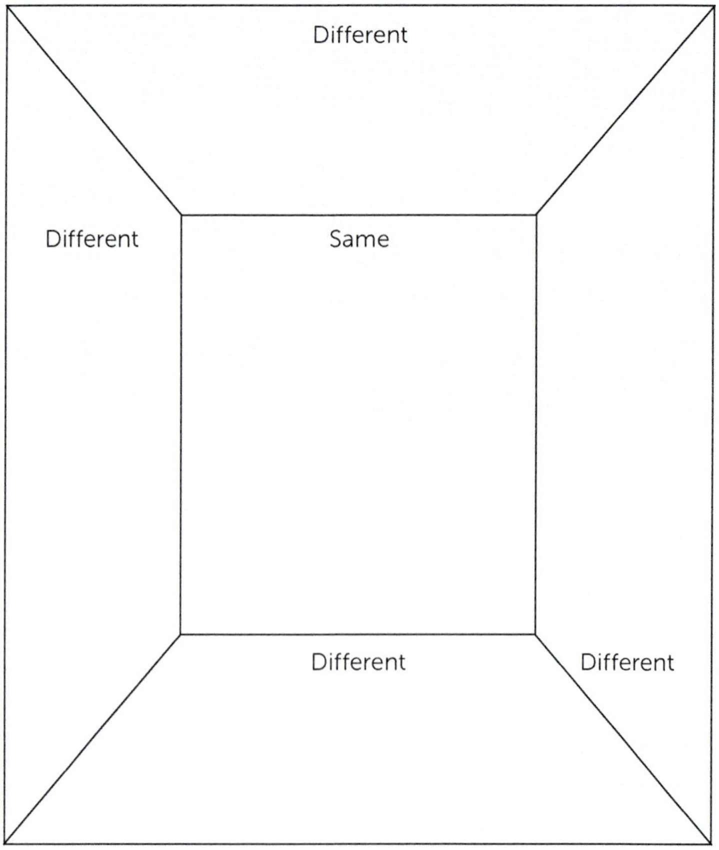

Reflecting on a Phenomenon

23 Finalize the anchor phenomenon response

Have students revisit their explanation of the pile of earth material in the schoolyard. Ask,

→ *What new information can you add to explain what caused this pile of earth material to form?*

Show the visual in the Anchor Phenomenon Resource Video 1 as needed to facilitate students' thinking. Review the driving question:

→ *What caused this pile of earth material to form?*

A completed response might include these ideas.

Claim: The new pile of sand (soil) came from the soil at the top of the hill. If the pile of earth materials had a fan shape, it might be a landform called an alluvial fan. This fan shape would be evidence that the earth material was eroded from the top of the hill by water and deposited when the water flow slowed or the slope decreased.

Students might also include:

There could be canyons formed when the earth material was pushed or eroded by water. This could be determined using the settling test conducted in vials with earth material and water. You could observe the properties of materials on the hill and in the pile.

Different-sized particles of soil would settle at different spots in the pile. Heavier pebbles won't move as much, smaller particles like clay would move farther than the pile of sand. Some humus might even float on the water and be deposited farther than the clay.

Have students complete their explanation. Ask students to identify any potential problems erosion and deposition might cause on their schoolyard.

24 View online tutorial: "Soil Formation" (optional)

For students who might benefit from more time with the concepts of soil composition and formation, have them engage with the tutorial. The tutorial through photos describes how soil composition depends on location, and reviews weathering and erosion of different sizes of earth material particles. Different soils from four locations are displayed in vials. Students are then asked to match soil profiles in vials to locations along the Rio Grande in Texas.

PART 3 Reducing Erosion Impact

Teacher Background

Engineering Storyline

LEARNING OBJECTIVE:

- **Students use a virtual modeling tool to develop possible solutions that reduce the impact of hazards from erosion and gather data to make evidence-based engineering decisions about the stability and change of human and natural systems.**

In Part 3, students use a simulation to explore variables and design experiments to collect data to answer specific questions. Here are some things they will find out to connect to the storyline:

- Humans cannot completely eliminate erosion and deposition, but they can install barriers and land improvements that reduce the impact of these natural earth processes.
- Effective engineering designs can reduce erosion so that landslides are less likely to occur, reducing the impact of natural hazards on humans.

Students use this modeling tool to provide evidence for engineering or policy decisions that will minimize the impact of erosion and deposition on natural systems and human communities.

FOCUS QUESTION

What Solutions Can We Design to Reduce the Impact of Erosion and Deposition on Human Structures?

The material eroded and transported by water, wind, and ice is the raw material of new landforms. When a stream slows down or dries up and can no longer carry its load, when the wind stops blowing, or when a glacier melts, the materials each carried are deposited.

Science and Engineering Practices

- Asking questions and defining problems
- Developing and using models
- Planning and carrying out investigations
- Analyzing and interpreting data
- Constructing explanations and designing solutions

Disciplinary Core Ideas

- ESS2.A: Earth materials and systems
- ESS2.E: Biogeology
- ESS3.B: Natural hazards
- ETS1.B: Developing possible solutions

Crosscutting Concepts

- Cause and effect
- Systems and system models
- Stability and change

Aa

Constraint   Criteria

Deposition is the process by which eroded earth materials settle out of the air, water, or ice. The amount of material carried by water depends primarily on the energy in the stream. Anything that increases the velocity of the water—an increase in slope or the amount of water, or a reduction in the friction of the water in contact with the riverbed due to a greater volume of water—increases the stream's capacity for carrying a load. Anything that decreases the velocity or volume of water—a decrease in slope with no increase in water volume, an obstruction such as a dam, a stream emptying into a larger body of water—will cause a stream to drop its burden as sediment. The larger, heavier sediments are deposited first, then the finer sand and silt.

Most earth-shaping events happen on such a large scale and in such inaccessible places that it is impossible for students to experience them directly. But models—small, simplified representations of the larger objects and events—are used by scientists to better comprehend the larger realities.

Some earth-shaping events happen abruptly, as in the case of landslides. These natural hazards result from excessive erosion. In this part, students set up virtual experiments using a computer simulation to investigate and compare techniques that can mitigate erosion and deposition, and to engineer solutions that reduce the impact of erosion and prevent landslides.

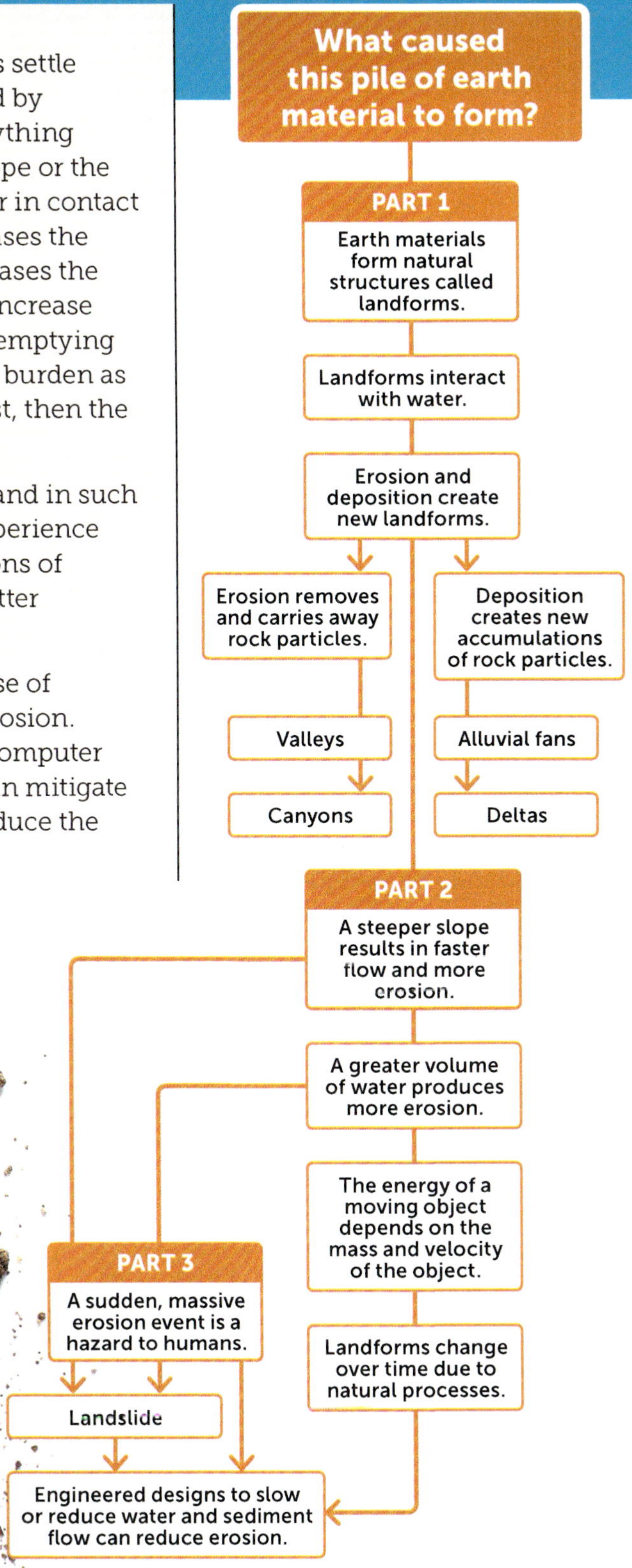

PART 3 Reducing Erosion Impact

Materials

FOSSWEB RESOURCES

For each student

- Notebook sheet 14, *Erosion Engineering Data*

For the class

- Online simulation: "Erosion Engineering"
- Teacher master 4, *Erosion Control Barriers*
- Word wall cards
- Teaching slides

For the teacher

- *Performance Assessment Checklist*
- *Assessment Record*

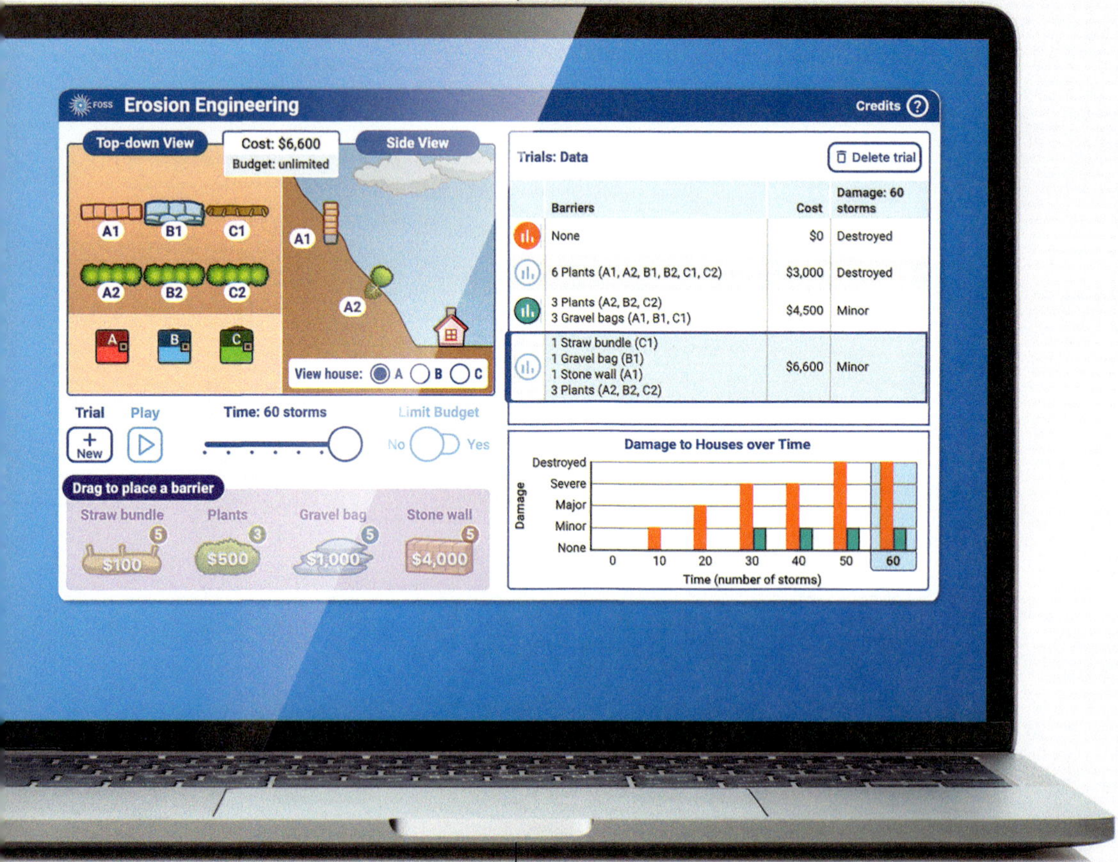

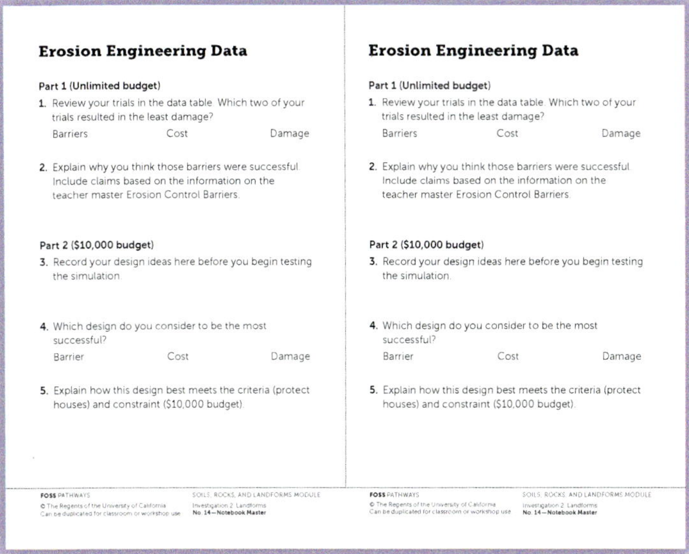

Erosion Engineering Data

Part 1 (Unlimited budget)

1. Review your trials in the data table. Which two of your trials resulted in the least damage?
   Barriers Cost Damage
2. Explain why you think those barriers were successful. Include claims based on the information on the teacher master Erosion Control Barriers.

Part 2 ($10,000 budget)

3. Record your design ideas here before you begin testing the simulation.
4. Which design do you consider to be the most successful?
   Barrier Cost Damage
5. Explain how this design best meets the criteria (protect houses) and constraint ($10,000 budget).

SOILS, ROCKS, AND LANDFORMS MODULE
Investigation 2: Landforms
No. 14—Notebook Master

Erosion Engineering Data

Part 1 (Unlimited budget)

1. Review your trials in the data table. Which two of your trials resulted in the least damage?
   Barriers Cost Damage
2. Explain why you think those barriers were successful. Include claims based on the information on the teacher master Erosion Control Barriers.

Part 2 ($10,000 budget)

3. Record your design ideas here before you begin testing the simulation.
4. Which design do you consider to be the most successful?
   Barrier Cost Damage
5. Explain how this design best meets the criteria (protect houses) and constraint ($10,000 budget).

FOSS PATHWAYS
© The Regents of the University of California
Can be duplicated for classroom or workshop use
SOILS, ROCKS, AND LANDFORMS MODULE
Investigation 2: Landforms
No. 14—Notebook Master

No. 14—Notebook Master

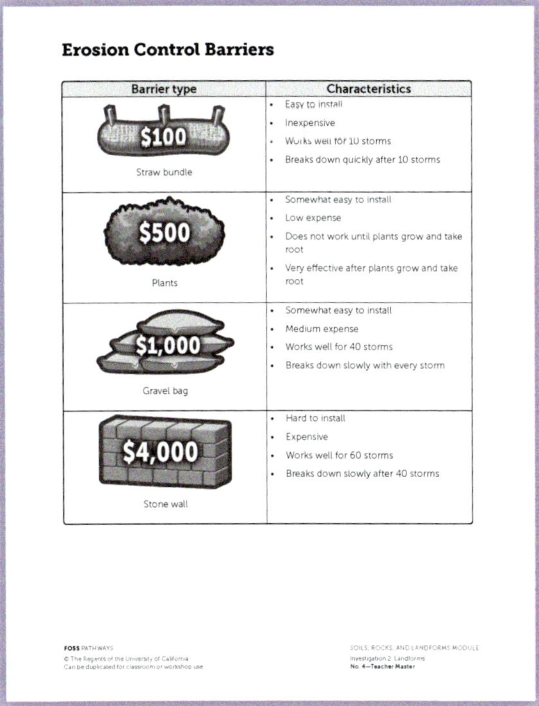

Erosion Control Barriers

Barrier type	Characteristics
$100 Straw bundle	• Easy to install • Inexpensive • Works well for 10 storms • Breaks down quickly after 10 storms
$500 Plants	• Somewhat easy to install • Low expense • Does not work until plants grow and take root • Very effective after plants grow and take root
$1,000 Gravel bag	• Somewhat easy to install • Medium expense • Works well for 40 storms • Breaks down slowly with every storm
$4,000 Stone wall	• Hard to install • Expensive • Works well for 60 storms • Breaks down slowly after 40 storms

SOILS, ROCKS, AND LANDFORMS MODULE
Investigation 2: Landforms
No. 4—Teacher Master

No. 4—Teacher Master

PART 3 Reducing Erosion Impact

Getting Ready

Schedule

4 Sessions
Session 1: Steps 1–12
Session 2: Steps 13–16
Session 3: Steps 17–20
Session 4: Steps 21–22 *I-Check 1–2*

Review

- Background for Part 3
- FOSSweb Resources for this part
- Online simulation: "Erosion Engineering"
- Teacher Preparation Video

Print or Copy

For each student:

- Notebook sheet 14, *Erosion Engineering Data*

For each group:

- Teacher master 4, *Erosion Control Barriers*

For the teacher:

- *Performance Assessment Checklist*
- *Assessment Record*

Prepare Material

- Plan for devices ______ A

Plan for Assessment

- Review Step 12, "What to Look For" in the performance assessment.
- Review Step 18 in Guiding to prepare for the I-Check. Also refer to the Assessment Chapter and Coding Guide for implementing and reviewing I-Checks, and using the online assessment system, FOSSmap.

PREPARATION DETAILS

A Plan for devices

Plan to have enough devices for students to work individually or in pairs. Test the functionality of "Erosion Engineering" on the students' devices, because the simulation is the source of data collection for the investigation.

The simulation builds on students' understanding of erosion and deposition developed using the physical stream tables in the previous two parts.

PART 3 Reducing Erosion Impact

Guiding the Investigation

ENGLISH LANGUAGE SUPPORT
Have students use the labeled diagrams and charts from the previous parts as a reference, if needed.

SESSION 1 45 MINUTES

1 Review erosion and deposition

Have students discuss what they've learned about the processes of erosion and deposition. If needed, use the questions to guide the discussion.

→ *What is erosion?* [The transportation of earth materials by water or wind.]

→ *What does it look like when erosion has happened?* [Valleys and stream channels are carved into the surface of the soil.]

→ *What is deposition?* [The settling of earth materials (sediment) in a new location by water or wind.]

→ *What does deposition look like?* [Earth-material particles.]

Students should discuss the stream tables and explain that erosion is the process that carries away sediment and that deposition is the settling of the materials in a new place. Students may also mention that slope and the amount of water that rushes over the land affects how much earth material is moved.

ENGLISH LANGUAGE SUPPORT
Show images or video clips of these events.

2 Describe the challenge

Ask students for examples of erosion and deposition that have caused changes to landforms and have impacted natural systems or human communities in their area. Have them consider floods, landslides or mudslides, rockfalls, or piles of sediment deposited in areas. Have students discuss the causes and effects of those events (damage to roads, bridges, or homes and property; clean up of surfaces).

Introduce the idea that people can plan for these possible events to reduce their impact. To do this, engineers and planners have to have data to make good decisions about what actions to take to prevent damage.

3 Introduce the focus question

FOCUS QUESTION

What solutions can we design to reduce the impact of erosion and deposition on human structures?

Write or project the focus question as you read it aloud.

→ *What solutions can we design to reduce the impact of erosion and deposition on human structures?*

Have students turn and talk to someone in their group about their ideas after they write the date and the focus question in their notebooks.

4 Describe the investigation

Tell students,

Today we will use a virtual simulation to model the effects of erosion and deposition. You will be asking questions and designing experiments to gather data. You will be acting as engineers to reduce erosion on a hillside and protect houses. If you don't protect the houses, the eroded soil will deposit on the houses and damage them. You could even have a landslide! Your ***criteria*** *for success is to have the least damage to the houses.*

5 Introduce the simulation

Display "Erosion Engineering" and walk students through the simulation. Start with the left side of the screen. Tell students the following as you point out each feature on the screen.

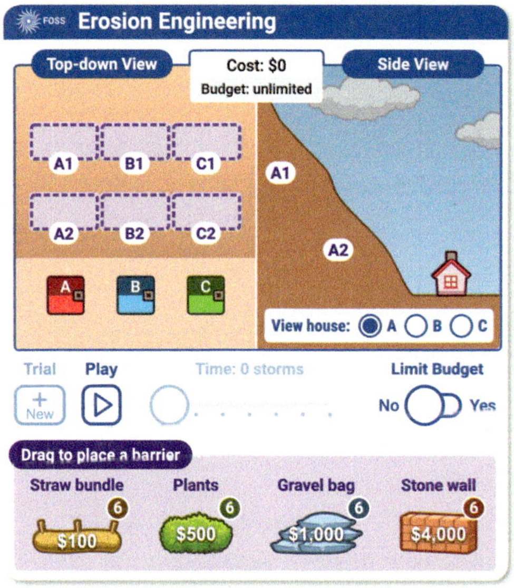

Start with "Top-down View" and "Side View."

In this simulation, you are trying to protect three houses, labeled A, B, and C, from erosion and deposition. You will be able to add barriers in two locations above each house. For example, location A1 and A2 protect house A. You can also see the side view for each house by selecting the house you want to see in the window.

Which barriers you use will be up to you to decide. We'll talk more about the barriers in a moment.

6 Introduce trials

Move on to the series of buttons and controls below the house view windows.

You will be testing a series of trials. The "+/New" button lets you begin a new trial. Each trial will generate a new row in the data table.

PART 3 Reducing Erosion Impact

SESSION 1

Point out to students the data table in the upper right, but don't explain the column headers yet.

For each trial, you will drag barriers of your choice up to the six locations above the houses. These barriers will help control erosion on the hill above a particular home.

Demonstrate how to drag a barrier to a slot above a home. Point out how the number of remaining barriers decreases when one is used. There are 6 barriers of each kind.

Note that each barrier has a cost labeled on it. We will start this simulation with an unlimited budget so you won't have to worry about cost. However, later I will present you with an engineering challenge, and you'll have to stick to a budget.

Show students the "Limit budget" toggle, and how turning it on and off affects the cost/budget window at the top left.

ENGLISH LANGUAGE SUPPORT
Make sure students understand the meaning of the word *barrier* and how it is used in this context.

7 Introduce barrier types

Project teacher master 4, *Erosion Control Barriers*. You can give each group, pair of students, or each student a copy of the teacher master as reference.

Read through the information with students, one barrier at a time. For each barrier, ask students to discuss in their group the pros and cons of using that barrier. Then ask a group to share their ideas with the class.

8 Discuss erosion data

Make sure the budget toggle is set to "No" and move some sample barriers above the houses. Then show students how to hit the "Play" button and observe results after 60 storms. Ask students to share how the Top-down view looks different, and show the side view for each house so students can observe what changed. Tell students,

This simulation models erosion and deposition on the hillside. Depending on which barriers you choose, you will experience more or less erosion. All the soil that erodes from the hill ends up depositing on the house! In some cases, if you do not protect the hillside from erosion, you may experience a landslide.

Ask students to share what they remember about landslides from earlier in the module. Confirm that a sudden, large amount of erosion is a landslide. Tell students that in this simulation, a landslide will destroy a home. They will use the simulation to find out which barriers create the best solutions to control erosion and protect the homes.

Point out that the results of this trial are automatically recorded on the data table in the upper right. They can replay the event by using the "Time: storms" slider controls to move back and forth in time.

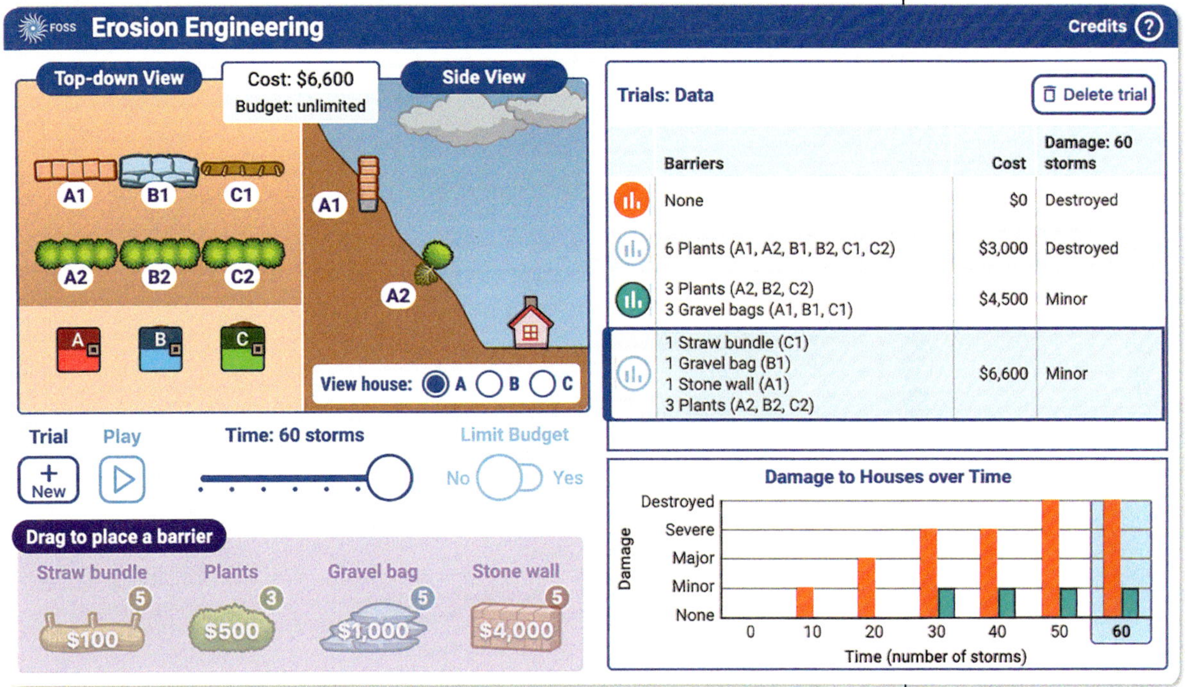

Students can add a trial to the bar graph by clicking on the graph icon in the data table for the row they want to add. This will select (or deselect) a trial to add to the bar graph on the lower right. Only two trials can be on the graph at once. The color for data on the bar graph matches the color of the selected data button in the table.

9 Focus the engineering challenge

Remind students that today they will work without a budget to test the erosion control barriers and learn more about their effectiveness. Leave teacher master 4, *Erosion Control Barriers*, up while students are working so they can think about the characteristics of each barrier. Suggest to students that they start out by testing only one kind of barrier in all six locations so they are able to compare the effectiveness of each barrier type. Any trials that they don't want to save they can delete with the "delete trial" button.

10 Distribute the notebook sheet

Distribute a copy of notebook sheet 14, *Erosion Engineering Data,* to each student. Explain that students will work on Part 1 today. Students should glue or tape the notebook sheet on a left notebook page so they can write on the blank right sheet. Students will want to write longer responses on the right page.

PART 3 Reducing Erosion Impact

SESSION 1

11 Model erosion in the simulation

Have students work at computers in pairs, or if necessary, in groups of four. Let them engage with the simulation at least 4–5 times. Visit students as they work and make sure they understand how to record their results.

12 Assess progress: performance assessment

Circulate from group to group throughout the investigation in this session and the next session. Conduct 30-second interviews, listen to group discussions and observe how students work together to prevent erosion. Make notes on the *Performance Assessment Checklist* while students work.

WHAT TO LOOK FOR

- *Students plan and carry out a variety of different trials to test design solutions.*
- *Students analyze and interpret the data in the table in order to compare results of the design solutions.*
- *Students use the graph tool to help compare results of the design solutions.*
- *Students construct explanations about how the characteristics of an erosion barrier may make it more or less effective in reducing erosion.*

Science and Engineering Practices

- Asking questions and defining problems
- Planning and carrying out investigations
- Analyzing and interpreting data
- Constructing explanations and designing solutions

Disciplinary Core Ideas

- ESS2.A: Earth materials and systems
- ESS2.E: Biogeology
- ESS3.B Natural hazards
- ETS1.B: Designing solutions to engineering problems

Crosscutting Concepts

- Cause and effect

SESSION 2 45 MINUTES

13 Discuss results

Start Session 2 with a review of the results. Use these questions to guide the discussion.

- ➜ *When you tested each type of barrier, which ones stopped the houses from being destroyed?* [Stone wall, gravel bags] *What data do you have to support this?* [The damage was only minor.]
- ➜ *Which barriers did not stop the houses from being destroyed?* [Plants and straw bundles] *What data do you have to support this?* [The damage was major.]
- ➜ *What pattern do you notice about the barriers?* [The walls and gravel bags stop the most, the straw only stops a little.]

Have volunteers share one of their most successful trials, and ask the students to explain why they think those barriers were successful. Give other students a chance to build on these ideas if they tried similar or different materials. Consider asking these questions.

- ➜ *What is your evidence that these barriers can prevent erosion?*
- ➜ *Did anyone experience a landslide? Under what conditions?*
- ➜ *How would having only $10,000 affect your choice of materials?*

Ask students to share any other observations. Students will analyze these barriers more in Step 17.

14 Introduce budget as a constraint

Tell students to work with the simulation again. Their criteria is the same: the least damage to the houses. But this time, they will have a **constraint**. They only have $10,000 to spend on the entire hillside and they should try to get the best protection from erosion spending the least amount of money. Slide the "Limit budget" toggle to "Yes," and point out the "Budget" area.

15 Review the notebook sheet

Have students read Part 2 of notebook sheet 14, *Erosion Engineering Data*. Give students 5 minutes to review the data they recorded in Part 1, discuss design ideas, and record their design ideas (question 3) before students begin to use the simulation.

16 Model erosion with a construction budget

Have students work at computers in pairs, or if necessary, in groups of four. Visit students as they engage with the simulation.

Stop the simulation when there are 10 minutes left in the session. Make sure students work on the last two questions on the notebook sheet before starting the sense-making discussion.

Science and Engineering Practices

- Analyzing and interpreting data

Crosscutting Concepts

- Systems and system models

PART 3 Reducing Erosion Impact

Constraint Criteria

ENGLISH LANGUAGE SUPPORT
If needed, post these questions and/or sentences started for students to use as reference.

Crosscutting Concepts
- Cause and effect
- Stability and change

NOTE
See the **Home/School Connection** for Investigation 2 at the end of the Differentiation Opportunities section. This is a good time to send it home with students.

SESSION 3 **45 MINUTES**

17 Review vocabulary

Review key vocabulary introduced in this part, criteria and constraint. If these words are not on the word wall, add them now.

18 Have a sense-making discussion

Have each group briefly share their most successful design. Students should answer these questions during their presentations.

→ *How does this design meet the criteria of protecting houses?*

→ *Did this design meet the constraint of staying within budget?*

→ *How did the data table and graph help you choose this design?*

Ask these more general questions and share their ideas.

→ *A common design is to put a straw or gravel barrier above a plant barrier. Why do you think this is?* [Plants take time to grow, and the barrier above would provide immediate protection for the home and the growing plants.]

→ *Did any of the barriers change over time?* [If you used straw in the first row and plants in the second, the straw got smaller over time, the plants grew. If you used just straw, they got smaller. If you used just plants, they washed away.]

→ *How did those changes affect the amount of erosion?* [Plants don't prevent much erosion until they grow. Straw will get carried away, so it isn't good for a long time.]

→ *Climate change increases the frequency of storms in many areas. How would this affect erosion?* [The straw, gravel, and stone barriers start to break down after a number of storms. If the storms happen more quickly, these barriers won't last as long.]

19 Respond to the focus question

Have students write a response in their notebooks.

→ *What solutions can we design to reduce the impact of erosion and deposition on human structures?*

Have students describe one of their designs and explain how it reduced erosion and deposition.

Reflecting on a Phenomenon

20 Review three-dimensions for Investigation 2

Distribute one or two self-stick notes to each student, and have them cut each note into three pieces, making sure each part has a sticky end.

Ask students to take 3 minutes to look back through their notebook entries to find the most important things they learned in Investigation 2. Students should include at least one science and engineering practice, one disciplinary core idea, and one crosscutting concept. They should tag those pages with self-stick notes.

When students have completed their notebook review and placed their tags, lead a short class discussion to create a list of three-dimensional statements that summarize what students have learned in this investigation. Here are the big ideas that should come forward in review discussions.

- We used stream tables to model erosion and deposition; these processes produce new and altered landforms, and contribute to soil formation.
- The data we collected showed that the size of earth material particles determines which particles settle out of water first (large particles), and which travel farther (small pieces).
- Explaining soil formation required us to think about weathering (Investigation 1) as well as erosion and deposition—pieces break off, then they are carried away and deposited elsewhere, often becoming part of the soil.
- Controlling the slope and the amount of water that ran through the stream table provided evidence that both variables affect the amount of erosion and deposition.
- We used a virtual model to compare engineering designs using erosion control barriers to reduce the impact of landslides on buildings.

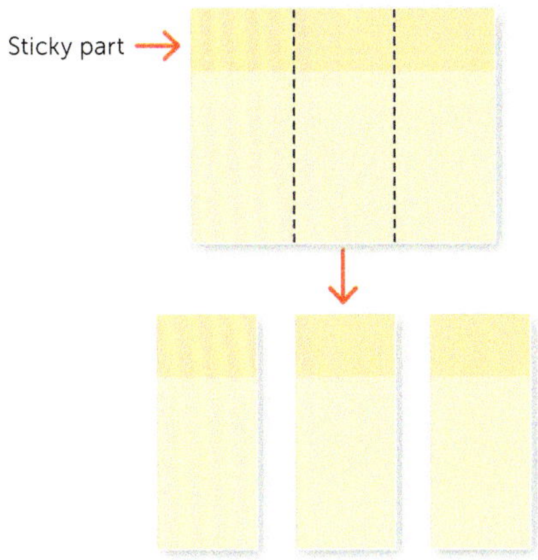

THREE-DIMENSIONAL LEARNING OBJECTIVES

- Students plan and carry out investigations using physical stream-table models to determine the effects of erosion and deposition on landforms. (Parts 1–2)
- Students observe, record, and analyze data to determine cause-and effect relationships about the rate and volume of erosion and the energy of moving water or wind. (Part 2)
- Students use a virtual modeling tool to develop possible solutions that reduce the impact of hazards from erosion and gather data to make evidence-based engineering decisions about the stability and change of human and natural systems. (Part 3)

PART 3 Reducing Erosion Impact

SESSION 4 **45 MINUTES**

21 Assess progress: I-Check

At the end of every one or two investigations, students take an I-Check benchmark assessment. I-Checks provide students with an opportunity to demonstrate their learning. When you return the I-Checks, students have another opportunity to reflect on their thinking by engaging in next-step strategies and confirm or revise their ideas.

Give the I-Check assessment one or two days after the final review. Distribute a copy of *Investigations 1–2 I-Check* to each student or schedule it on FOSSmap for students to take online.

See the Assessment chapter for guidance and information about using the online assessment system, FOSSmap. You can schedule the I-Check on FOSSmap and then access data to look carefully at students' progress.

NOTE
Plan on 30 minutes for students to take the assessment and a 15-minute session a day or two later for a next-step strategy.

22 Use a next-step strategy

Review student responses using the information found in the Assessment Coding Guides on FOSSweb. As you review the items, note any concepts students may be struggling with. Plan for a next-step strategy session focusing on the disciplinary core ideas, science and engineering practices, and crosscutting concepts students may need additional help with. Refer to the Assessment chapter for more information about next-step strategies.

I-Checks are valuable when you use them as formative assessments. To do this, code students' tests, but do not put any marks on them. Thinking about the content of the questions comes to a halt when students see marks on their papers. Using self-assessment next-step strategies is essential to develop a class culture focused on a growth mindset that values assessment as a learning process.

NOTE
During or after these next-steps with the I-Check, you might ask students to make choices for possible derivative products based on their notebooks for inclusion in a summative portfolio. See the Assessment chapter for more information about creating and evaluating portfolios.

Differentiation Opportunities

Below is a sample of student-choice activities related to this investigation for group or individual projects. Students can select an activity that is of special interest to them, or the class can choose a project for collaboration. Extensions and project ideas are written for students and accessible on FOSSweb.

Language Extensions

- Research erosion in a region of your state, the nation, or the world.
- Research the negative and positive effects of flooding in a specific natural area.

Science Extensions

- Go on a neighborhood walk after a hard rain to observe erosion and deposition.
- Investigate local soil erosion.
- Design and carry out a new stream-table investigation.

Math Extension

- Analyze data on the effect of slope on erosion.

Environmental Literacy Extension

- Interview a local erosion-control expert, soil scientist, or geologist.

Home/School Connection

Students explore media and the Internet for news about landforms and the processes that create and change them. They prepare a report that includes the details of the event, where it happened, what types of landforms were described, and how the landforms were created or changed by natural or human-caused processes.

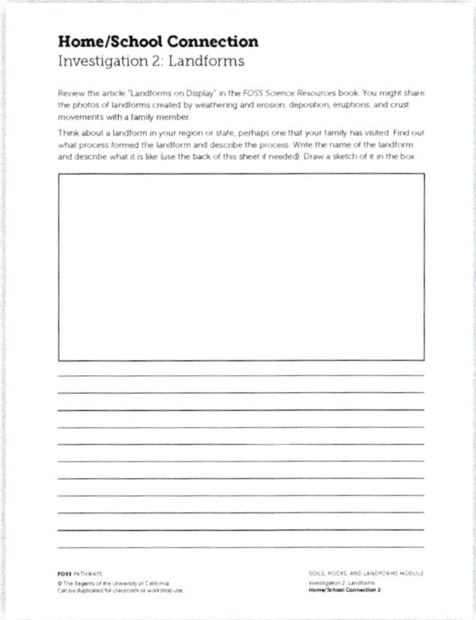

Home/School Connection
Investigation 2: Landforms

Review the article "Landforms on Display" in the *FOSS Science Resources* book. You might share the photos of landforms created by weathering and erosion, deposition, eruptions, and crust movements with a family member.

Think about a landform in your region or state, perhaps one that your family has visited. Find out what process formed the landform and describe the process. Write the name of the landform and describe what it is like (use the back of this sheet if needed). Draw a sketch of it in the box.

FOSS PATHWAYS
© The Regents of the University of California
Can be duplicated for classroom or workshop use.

SOILS, ROCKS, AND LANDFORMS MODULE
Investigation 2: Landforms
Home/School Connection 2

No. 2—Home/School Connection

INVESTIGATION 3

Mapping Earth's Surface

INVESTIGATION 3

Mapping Earth's Surface

Phenomenon Storyline

Hiking to the peak of a mountain, Part 1: A group of hikers have to solve a problem. They want to hike to the top of a volcanic peak and have selected a mountain in the Cascade Mountains for this adventure. They have a poster of the mountain and have been studying it, but they need more information about the terrain and elevations to decide on a hiking route.

What kind of a map do the hikers need and how will it help them to plan a route to the mountain peak?

Impact to Mount St. Helens, Part 2: Students were looking at maps and images of the same volcanic mountain at different times. They observed some major differences in the landforms.

How did the mountain change and what caused the change?

Students focus on landforms that rise above their surroundings by representing landforms in three dimensions using topographic maps, photographs, and physical models. Two volcanic mountains—Mount Shasta and Mount St. Helens—are the major landforms that students investigate as they work with models to represent and describe mountain surface features and understand how mountains were formed and change.

THREE-DIMENSIONAL LEARNING OBJECTIVES

- Students ask questions and use maps and scale models to gather data about mountain landforms using scale, proportion, and quantity. **(Parts 1–2)**
- Students collect, analyze, and interpret data to describe stability and change to a mountain after a volcanic eruption. **(Part 2)**

INVESTIGATION 3

Mapping Earth's Surface

At a Glance

Investigation Part	Sessions	Practices and Crosscutting Concepts
Introduce problem—Hiking to the peak of a mountain **PART 1 Using Models and Topographic Maps** **FOCUS QUESTION: How can we represent the different elevations of a mountain?** Students build and use a model of Mount Shasta by stacking and orienting six foam layers. They trace outlines of the six pieces onto paper, creating a topographic map of the mountain. With this foundation, students use a USGS topographic map to solve the problem about a path to the peak of a volcanic mountain.	**SESSION 1*** Steps 1–11 **SESSION 2** Steps 12–20 **SESSION 3** Steps 21–27 (Steps 22, 28 optional) **SIDE TRIP 3,** after Part 1 *A session is 45 minutes.	**Practices** • Asking questions • Developing and using models • Using mathematics and computational thinking • Analyzing and interpreting data • Constructing explanations • Obtaining, evaluating, and communicating information **Crosscutting Concepts** • Scale, proportion, and quantity
Introduce anchor phenomenon 2—Impact to Mount St. Helens **PART 2 Mount St. Helens Case Study** **FOCUS QUESTION: How did the mountain change and what caused the change?** Students compare two topographic maps and have a short debate about whether or not they show the same mountain. After learning that the two topographic maps are the same mountain, they gather information about Mount St. Helens before and after its devastating eruption in 1980. Students watch a USGS video that explains how scientists were involved in predicting the eruption. Students describe how the loss of stability caused major changes to the landform.	**SESSION 1** Steps 1–14 (Steps 9, 11–12 optional)	**Practices** • Developing and using models • Analyzing and interpreting data • Constructing explanations • Engaging in argument from evidence • Obtaining, evaluating, and communicating information **Crosscutting Concepts** • Scale, proportion, and quantity • Stability and change

Performance Expectations

4-ESS2-2. Analyze and interpret data from maps to describe patterns of Earth's features.

4-ESS3-2. Generate and compare multiple solutions to reduce the impacts of natural Earth processes on humans.

Disciplinary Core Ideas	Interpreting Information	Assessment
ESS2.B: Plate tectonics and large-scale system interactions • A topographic map uses contour lines to show the shape and elevation of the land. • The change in elevation between two adjacent contour lines is always uniform. • The closer the contour lines, the steeper the slope and vice versa.	**Science Resource Book** "Topographic Maps" (optional) "A Cascades Volcano: Mount Shasta" **Other Media** Mount Shasta Avalanche Center (climbing maps and photos) Online activity: "Topographer" (optional Side Trip 3)	**Embedded Assessment** Response sheet **NGSS Performance Expectation addressed** 4-ESS2-2
ESS2.B: Plate tectonics and large-scale system interactions **ESS3.B:** Natural hazards • The surface of Earth is constantly changing; sometimes those changes take a long time to occur, and sometimes they happen rapidly. • Catastrophic events have the potential to change Earth's surface quickly. • Scientists monitor volcanoes as they erupt to learn more about the earth-changing events.	**Science Resource Book** "What Geoscientists Do" (optional) **Other Media** Video: *Mount St. Helens' Impact* Anchor Phenomenon Resource Video 2	**Embedded Assessment** Performance assessment **NGSS Performance Expectations addressed** 4-ESS2-2 4-ESS3-2

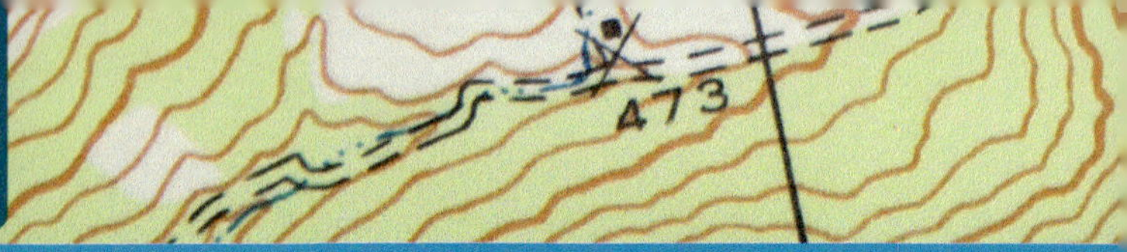

INVESTIGATION 3

PART 1 Using Models and Topographic Maps

Science and Engineering Practices

- Asking questions
- Developing and using models
- Using mathematics and computational thinking
- Analyzing and interpreting data
- Constructing explanations
- Obtaining, evaluating, and communicating information

Disciplinary Core Ideas

- ESS2.B: Plate tectonics and large-scale system interactions

Crosscutting Concepts

- Scale, proportion, and quantity

Aa

contour interval
contour line
elevation
satellite cone
sea level
topographic map
volcano

Teacher Background

Phenomenon Storyline

LEARNING OBJECTIVE:

- **Students ask questions and use maps and scale models to gather data about mountain landforms using scale, proportion, and quantity.**

The scenario for this first part introduces a common problem, how to get from one place to another over difficult terrain. A group of hikers want to hike to the top of a volcanic peak and have selected a mountain in the Cascade Mountains for this adventure. They have a poster of the mountain and have been studying it, but they need more information about the terrain and elevations to decide on a hiking route.

What kind of a map do the hikers need and how will it help them to plan a route to the mountain peak?

Part 1 begins the use of representation tools to describe the structure of a volcanic mountain. Here are some things students will find out to connect to the storyline.

- Mountains have different features and elevations.
- Maps and models showing changes in elevation can be used to make decisions.
- A topographic map uses contour lines to show the shape and elevation of the land.
- A physical model is a scale model showing the shape and features of the mountain and can be used to generate a topographic map.

FOCUS QUESTION

How Can We Represent the Different Elevations of a Mountain?

There's something about the landform we call a mountain that compels people to climb to the top. Mount Everest, Mount Whitney, the Matterhorn—their names and locations are familiar to many, and standing on their peaks is the goal of outdoor adventurers. Mount Shasta and Mount St. Helens are also mountains of great height. Mount Shasta (14,180 feet; 4,322 meters) rises impressively above most of the landforms in northern California. Mount St. Helens (8,336 feet; 2,550 meters) is in the state of Washington. Both mountains are volcanoes in the Cascade Range. Mount Shasta is inactive, but Mount St. Helens has been quite active during the last 45 years. The view that we most often get of a mountain like Mount Shasta or Mount

St. Helens is the profile. Another useful view of a mountain is from above. Topographic maps are two-dimensional representations of three-dimensional surfaces as seen from directly overhead. Topography is the shape of the surface of the land, its size, relief, and elevation. Mountains, plains, canyons, and cliffs are a few of the topographic features that contribute to the diversity of Earth's surface.

Contour lines are characteristic of topographic maps. Any one contour line represents a specific elevation, which is the vertical distance above sea level. Sea level is zero elevation; it is defined as the average distance between the highest high tide and the lowest low tide around the world.

When you trace the 9,000-foot contour line on a topographic map, everything that lies on that line is exactly 9,000 feet above sea level. Contour lines show the contour, or shape, of the land at one specific elevation.

When you look at a topographic map, you will notice that some of the contour lines are relatively close together, while others are quite far apart. Lines close together indicate regions with steep slopes. As you move from one line to another, the change in elevation remains constant, whether you move up or down. The vertical distance between contour lines is the contour interval.

The US Geological Survey (USGS) has been producing a standard series of topographic maps since 1882 as part of its National Mapping Program. In 2009, the USGS began the release of a new generation of digital topographic maps that are available online.

NOTE: USGS maps, which students will use in this investigation, are measured in feet. The scale of the map is 1:24,000 (1 inch: 24,000 inches or 2,000 feet). The USGS produced most of the topographic maps in this country in feet and has produced a few maps in metric. The area portrayed by the 7.5-minute series ranges from 49 to 64 square miles. It takes about 57,000 7.5-minute maps to cover the 48 contiguous states.

The conceptual flow for this investigation starts as students are presented with a problem of planning a hike to the top of a mountain. They are introduced to topography by building a model of a landform—a mountain. They use a foam model of Mount Shasta to create a useful tool—a topographic map. They are introduced to elevation above sea level and the conventions of contour lines and contour intervals. Students develop and use this model to gather information about the mountain and the best way to hike to the top.

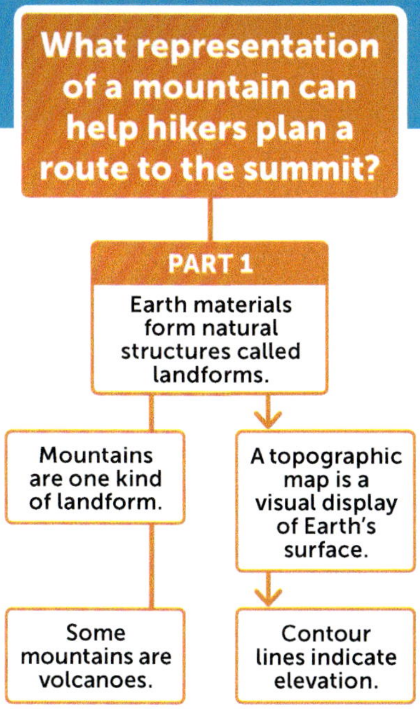

INVESTIGATION 3

PART 1 Using Models and Topographic Maps

Materials

PROVIDED EQUIPMENT

For each student

1 *FOSS Science Resources: Soils, Rocks, and Landforms*
"Topographic Maps" (optional)
"A Cascades Volcano: Mount Shasta"

For each group

1 Foam mountain set, 6 pieces, and dowel in large zip bag

1 Topographic map of Mount Shasta, CA (USGS 7.5-minute quadrangle, scale 1 inch=2,000 feet)

1 *Foam-Mountain Topographic Map* transparency (same as teacher master 5 in a transparent overly)

For the class

1 Poster, *Mount Shasta*

SCHOOL-SUPPLIED ITEMS

For each group

4 Pencils

FOSSWEB RESOURCES

For each student

- Teacher master 6, *Mountain Map*
- Notebook sheet 15, *Response Sheet A—Investigation 3*
- Notebook sheet 16, *Response Sheet B—Investigation 3*

For each group

- Teacher master 5, *Foam-Mountain Topographic Map*

For the class

- Word wall cards and equipment cards
- Teaching slides

For the teacher

- *Embedded Assessment Notes*

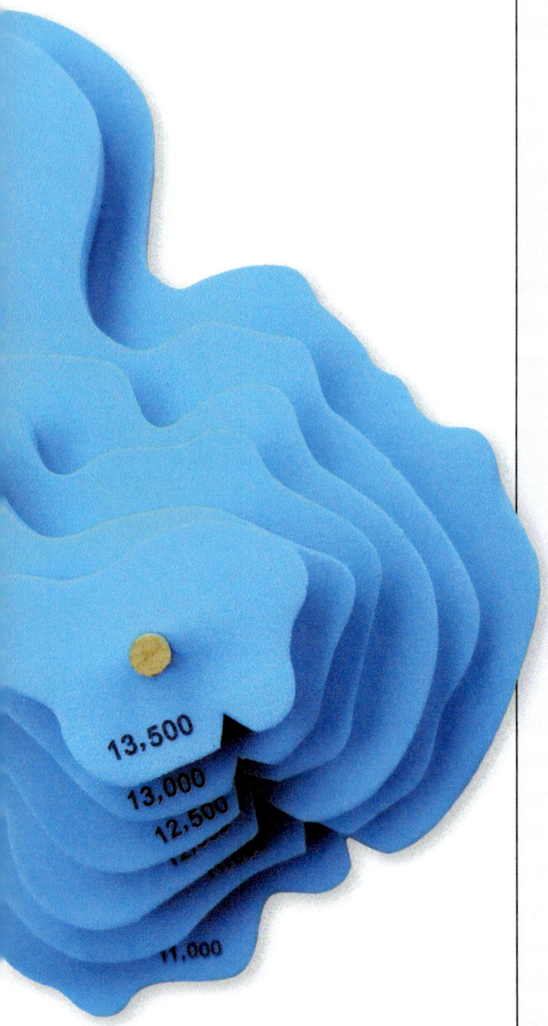

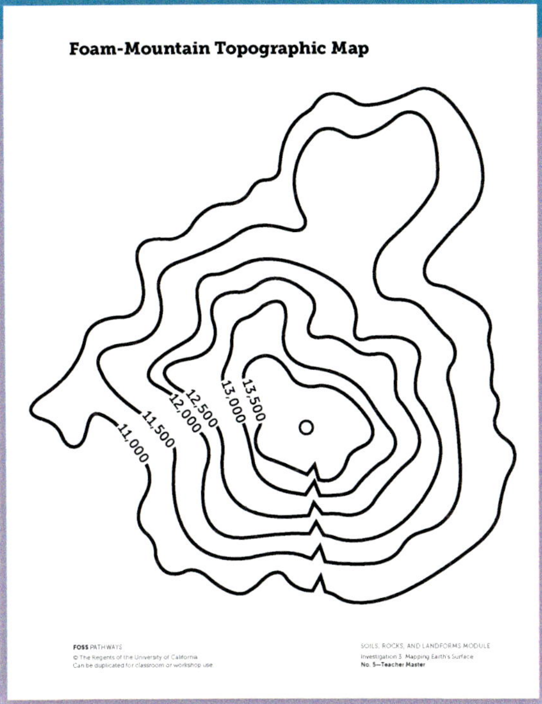

No. 5—Teacher Master

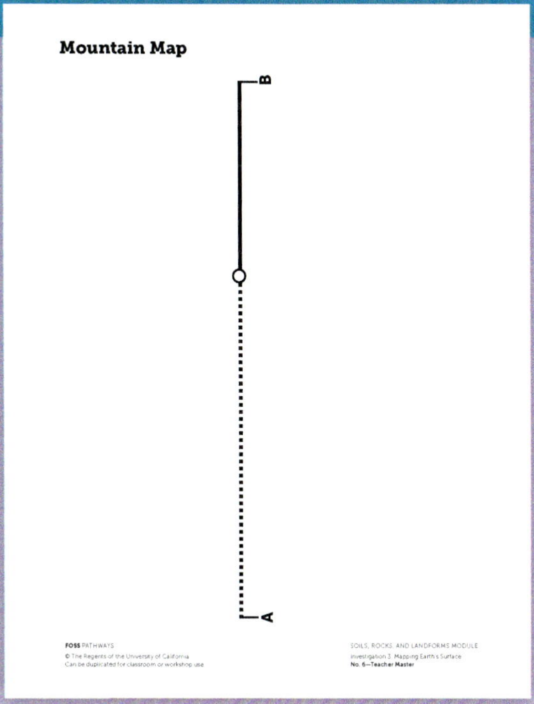

No. 6—Teacher Master

Response Sheet A—Investigation 3

William and his uncle are planning a hike to Mallard Peak, a landform in a nearby park. They will park the car near the picnic area close to the highway and hike from there.

They have a topographic map of the area. They want to use it to plan the best route to the peak. They don't mind climbing up steep slopes and want to stop by a waterfall on the way up. They think it would be good to walk down a less steep slope at the end of the day.

Help William and his uncle out. Use the map on *Response Sheet B* to draw a trail that you suggest they take to get to the top of the peak and another trail down. Label the start and end points. Use arrows to show the direction they should hike.

On a blank page in your notebook, explain why you think the route you drew will be the best one.

FOSS PATHWAYS
© The Regents of the University of California
Can be duplicated for classroom or workshop use

SOILS, ROCKS, AND LANDFORMS MODULE
Investigation 3: Mapping Earth's Surface
No. 15—Notebook Master

Response Sheet A—Investigation 3

William and his uncle are planning a hike to Mallard Peak, a landform in a nearby park. They will park the car near the picnic area close to the highway and hike from there.

They have a topographic map of the area. They want to use it to plan the best route to the peak. They don't mind climbing up steep slopes and want to stop by a waterfall on the way up. They think it would be good to walk down a less steep slope at the end of the day.

Help William and his uncle out. Use the map on *Response Sheet B* to draw a trail that you suggest they take to get to the top of the peak and another trail down. Label the start and end points. Use arrows to show the direction they should hike.

On a blank page in your notebook, explain why you think the route you drew will be the best one.

FOSS PATHWAYS
© The Regents of the University of California
Can be duplicated for classroom or workshop use

SOILS, ROCKS, AND LANDFORMS MODULE
Investigation 3: Mapping Earth's Surface

No. 15—Notebook Master

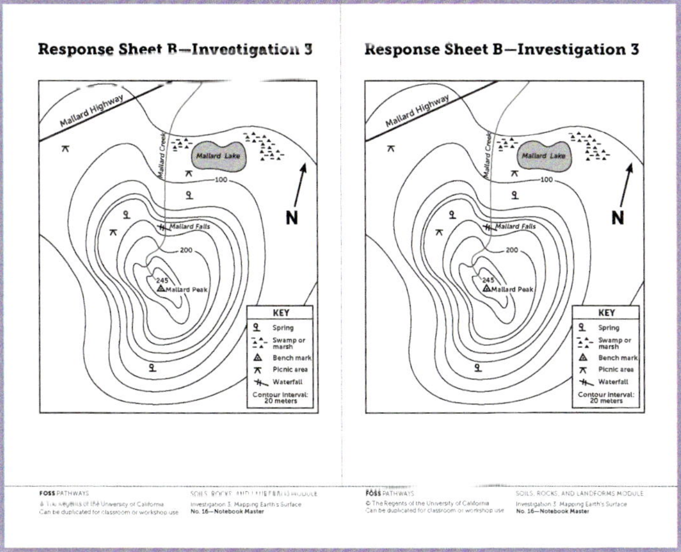

No. 16—Notebook Master

PART 1 Using Models and Topographic Maps

Getting Ready

Schedule

3 Sessions
Session 1: Steps 1–11
Session 2: Steps 12–20
Session 3: Steps 21–27 (Steps 22 and 28 optional)

Review

- Background for Part 1
- FOSSweb Resources for this part
- Optional reading: "Topographic Maps"
- Reading: "A Cascades Volcano: Mount Shasta"
- Online resources: Mount Shasta Center (climbing maps and photos)
- Teacher Preparation Video

Print or Copy

For each student:

- Teacher master 6, *Mountain Map* (make a few extras for the class)
- Notebook sheet 15, *Response Sheet A—Investigation 3*
- Notebook sheet 16, *Response Sheet B—Investigation 3*

For each group:

- Teacher master 5, *Foam-Mountain Topographic Map*

For the teacher:

- *Embedded Assessment Notes*

Prepare Material

- Check foam mountains A
- Practice making a mountain map B

Plan for Assessment

- Review Step 21, "What to Look For" in the response sheet. Use *Response Sheet—Investigation 3* for a closer look at students' ability to gather information from a topographic map and how to construct an explanation for the route planning. Plan to spend time having students reflect on their answers after you have reviewed them.

PREPARATION DETAILS

A Check foam mountains

Each collaborative group will build a model mountain by stacking six foam layers on a dowel. Each layer has a center hole for the dowel and a notch for alignment. Check that each bag has 6 model pieces and a dowel.

Assemble one model mountain for the introduction. It is very important to make sure each piece is correctly oriented and right side up. Refer to teacher sheet 5, *Foam-Mountain Topographic Map*. When the mountain is assembled, the elevations can be read.

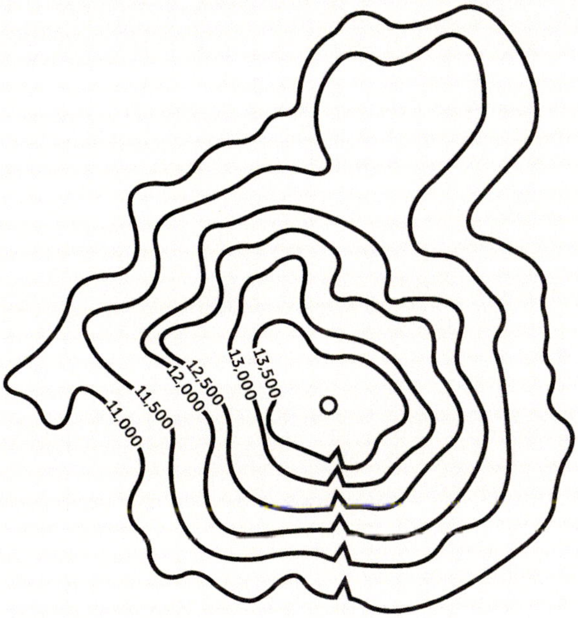

NOTE

The topographic maps in this part show elevation in feet because the USGS maps for Mount Shasta are in feet.

B Practice making a mountain map

Practice making your own topographic map by following Step 10 in Guiding the Investigation.

Check your *Mountain Map* by laying a *Foam-Mountain Topographic Map* transparency on top of the map you drew. If you lined up the notches and traced carefully around the foam pieces, the two maps should match. (Make sure you have the transparency right side up.)

Guiding the Investigation

Science and Engineering Practices

- Asking questions

ENGLISH LANGUAGE SUPPORT
Use the poster as a reference as you explain the concept of elevation.

SESSION 1 **45 MINUTES**

1 Introduce the problem to be solved

Introduce the problem.

A group of hikers have to solve a problem. They want to hike to the top of a volcanic peak and have selected a mountain in the Cascade Mountains for this adventure. They have a poster of the mountain and have been studying it, but they need more information about the terrain and elevations to decide on a hiking route.

Show them the *Mount Shasta* poster and tell them the name of the mountain. Mount Shasta is a **volcano** in the Cascade Mountains in the western United States.

Give students time to observe the poster and to ask questions about the mountain's features (how high is it, how steep is it, is there snow on top, etc.). Then introduce the driving question.

What kind of a map do the hikers need and how will it help them to plan a route to the mountain peak?

2 Review stream-table maps

Ask students to recall the stream-table drawings they made when they recorded their observations of the stream-table models. These representations are a kind of map. Have them turn to one of the stream-table observation maps in their notebooks. Have them discuss in their groups the kinds of information those maps provided.

➜ *What landforms were you able to map on your stream-table observation sheet?*

Call for attention and have a few students share the different landforms they recorded on their maps (canyons, valleys, rivers, and alluvial fans).

3 Discuss map limitations

Weave into the discussion the idea that the stream-table map shows the shape of the different landforms, but it does not provide any information about three-dimensional aspects. For example, there is no indication of how deep the canyon is.

Tell students that when we talk about vertical distance (going up or down in a canyon, mountain, or other landform), we are talking about **elevation**. Elevation is the vertical distance or height above (or below) **sea level**.

4 Introduce the focus question

FOCUS QUESTION

How can we represent the different elevations of a mountain?

Say aloud the focus question as you project or write it on the board. Have students write the date and focus question in their notebooks.

➜ *How can we represent the different elevations of a mountain?*

5 Demonstrate how to build a model

Suggest one way to find the answer to some of the questions is to make a three-dimensional model of the mountain. (Ask students what the three dimensions are [length, width, and height].) Show students a set of six foam pieces. Point out that the pieces are all different sizes and shapes, but each is the same thickness. Point out the numbers on the layers; the numbers should remain facing up.

Demonstrate how to punch the dowel up through the hole near the center of each foam piece. Tell them to start with the piece labeled with the lowest number (11,000), then the next lowest (11,500), continuing until all six are in place. Show them how to align the notches.

Have Getters pick up the materials and start building models.

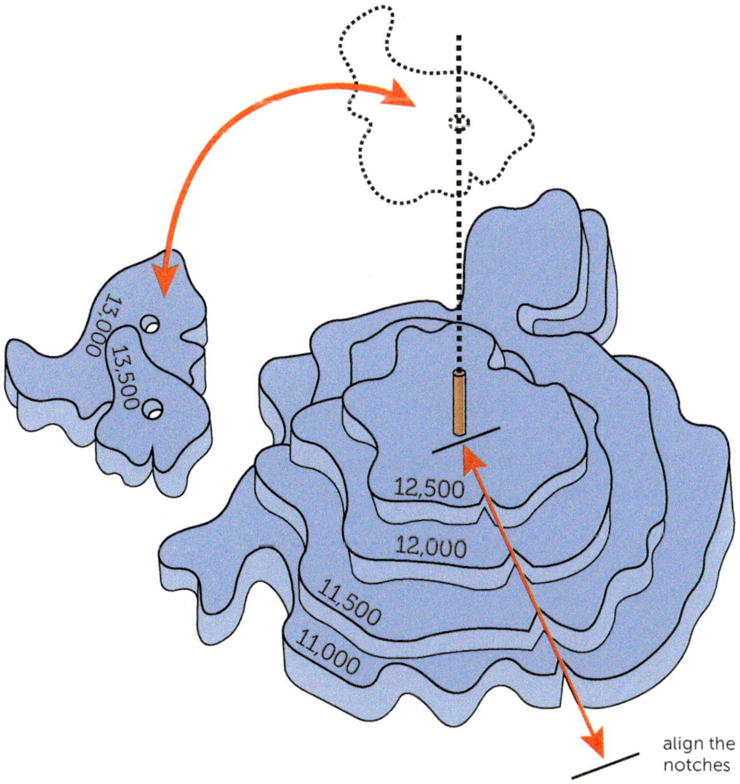

NOTE

Encourage students to push the dowel through the holes from the bottom of the foam pieces.

INVESTIGATION 3

PART 1 Using Models and Topographic Maps

Science and Engineering Practices

- Developing and using models

Crosscutting Concepts

- Scale, proportion, and quantity

NOTE

Keep the poster of Mount Shasta displayed on the wall throughout this investigation.

SESSION 1

6 Describe the model landform

Explain that the scale model is smaller than the actual mountain but represents its shape at different levels. Have students compare the foam mountain to the real mountain on the *Mount Shasta* poster. Ask,

→ *How does our model mountain compare to a real mountain?* [Most mountains aren't built like stair steps, but have smoother slopes; real mountains are much bigger; they aren't blue; and so forth.]

Point out the two pieces of the mountain that extend way beyond the other pieces at the 11,000- and 11,500-foot levels. These layers represent the **satellite cone** of Mount Shasta called Shastina. Explain that a satellite cone forms on the flank of a volcano where a crack allows accumulation of volcanic materials beyond the central vent. Encourage students to find Shastina on the poster. Say,

Shastina is 12,329 feet high. Were it considered a separate mountain, it would be the third highest in the Cascade Range! If we wanted to make the foam-mountain model more accurate, we could add another foam piece to show Shastina at the 12,000-foot level.

7 Circle the mountain

Draw students' attention to the numbers printed on the top edges of the layers. Explain that each layer is labeled with an elevation above sea level. For example, the top of the first layer represents 11,000 feet. Trace around the top edge of the first layer as you say,

If you travel all the way around the mountain on top of this layer, you will always be 11,000 feet above sea level.

Have students take finger walks around different layers of the foam mountains and identify their elevations.

8 Discuss equal intervals

Ask students to determine the difference in the elevation between the top of the first layer (11,000) and the top of the second layer (11,500).

→ *Is the vertical distance from the top of any one layer of foam to the top of the next layer always the same?*

Confirm that, because the layers are the same thickness, the vertical distance from the top surface of any layer to the top of the next layer is always the same; in this case, 500 feet.

9 Introduce *topographic map*

Tell students that you would like to make a flat map that would show both the shape of the landform and the elevation at the same time. A map would be easier to carry on a hike than a model. Have students discuss ideas in their groups for a few minutes. Share their ideas and guide the discussion to topographic maps.

There is a kind of map that shows both the elevation and the shape of the land. It is called a ***topographic map*** *and is another kind of model. You can use the foam pieces to draw a topographic map of your foam model mountain. Both models are scale models—a detailed representation of something that is much larger.*

10 Demonstrate the mapmaking procedure

Explain as you demonstrate the procedure.

a. Take the foam mountain apart.

b. Poke a pencil point from the top through the hole of one of the layers; make sure the elevation label is showing (on the top side).

c. Set the point of the pencil on the center circle of the *Mountain Map*.

d. Line up the notch with the solid dark line that runs from the circle to point B.

e. Hold the foam piece securely, and carefully trace around its outside edge.

f. Repeat this procedure for all six pieces of foam. The order in which the pieces are traced doesn't matter, as long as you always line up the hole in the foam mountain with the circle and the notch with the solid black line.

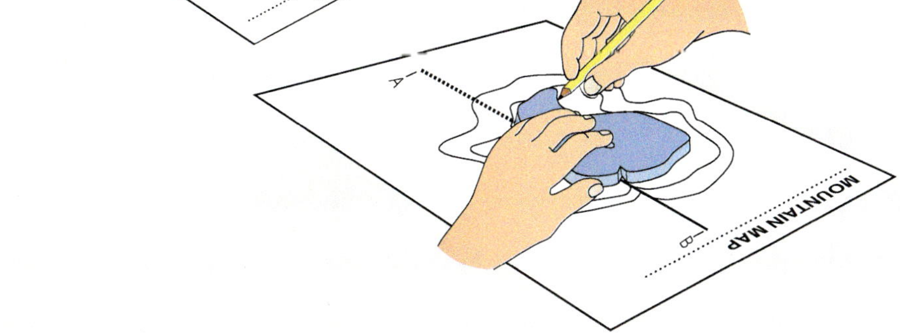

Crosscutting Concepts

- Scale, proportion, and quantity

NOTE

Students should use pencils, not pens, to trace the foam pieces.

PART 1 Using Models and Topographic Maps

SESSION 1

11 Draw the topographic map

Have Getters get four copies of teacher master 6, *Mountain Map*, and have students start tracing the foam pieces. When they finish, ask the groups to reassemble the model.

As students begin to finish their maps, distribute a *Foam-Mountain Topographic Map* transparency to each group. Students can use this transparency to check their work. If they lay the transparency over their map (they must be able to read the transparency title right side up), the lines students drew should closely match the lines on the transparency. Point out to students that the peak of Shastina is indicated on the transparency. They can mark the approximate location of Shastina on their topographic maps.

NOTE

Monitor students' progress as they work, making sure they have the proper alignment and side face up for the foam pieces. You might want to stop work briefly after about 5 minutes to do a quick check.

MATERIALS FOR STEP 11

- **Mountain Map** sheets
- **Foam-Mountain Topographic Map** transparencies

SESSION 2 45 MINUTES

12 Compare maps and mountains

Ask students to compare their topographic map with the view of the foam-mountain model looking down from directly overhead on the top of the mountain. Have them locate the highest and lowest areas on their topographic maps.

Science and Engineering Practices

- Developing and using models
- Using mathematics and computational thinking

13 Introduce contour lines

Explain that the lines they have traced on the paper are called **contour lines**. Tell them that contour lines are imaginary lines that follow the surface of the mountain at a precise elevation. For example, wherever the 11,000-foot contour line is, the elevation of the land is 11,000 feet.

14 Introduce contour interval

Remind students that the thickness, or height, of each piece of foam is the same. On topographic maps, the difference in elevation from one contour line to the next is always the same, too. This distance is the **contour interval**. Ask,

→ *What is the contour interval on your topographic map?* [500 feet]

Set the foam mountains off to the side, on a countertop or other location close by, as they will be used again in Step 16.

NOTE

Emphasize that contour lines are not always the same distance apart on the map, but they always represent equal steps in elevation.

15 Compare topographic maps

Tell students you have another topographic map for them to use. This one was designed by the United States Geological Survey. Distribute one USGS topographic map of Mount Shasta to each group. Let them get oriented to the map, looking at the colors, the labels of the landform features, and the contours.

Challenge the groups to position the *Foam-Mountain Topographic Map* transparency to match the USGS map. If students have difficulty doing this, have them locate the peaks of Mount Shasta and Shastina on the topographic map. (Each is marked with a + sign, the name of the peak, and its elevation.) Once students have located those two peaks, they should be able to orient the transparency on the map and see how the 11,000 contour line matches. They might also want to place the foam mountain on the map. Here are some guiding questions.

→ *What is the contour interval of the USGS map?* [40 feet]

→ *How were you able to match the topographic map transparency to the USGS map?* [Follow the 11,000-foot contour interval]

→ *Can you match the foam mountain to the USGS map?*

→ *Can you find glaciers and lava flows on the map?*

→ *What questions do you have about the topographic map?*

Have students label the elevations of the peaks of Mount Shasta based from the benchmarks on the map (14,162 ft/4,317 m) and Shastina (12,330 ft/3,758 m) on the topographic maps they drew. (You may find slightly different elevations for the mountain peaks in different references.) Have students discuss this question.

→ *How do topographic maps help people "see" the shape of the land?*

16 Observe the slope

Ask students to look at the topographic maps they drew and ask,

→ *Which is the steeper side of the foam mountain and how do you know?* [Side B, the side with the notches; the lines are closer together]

Have students retrieve the foam mountain models and confirm that B is the steeper side. Have students look at the foam mountain from directly overhead. The contour lines based on the edges of the foam pieces will be closer together on the steeper side of the mountain.

Remind students that the contour interval is the same whether the contour lines are close together or far apart. Contour lines that are close together represent a steep slope. On the more gently sloped areas of the mountain, the contour lines are farther apart.

MATERIALS FOR STEP 15

- Topographic maps of Mount Shasta (USGS)
- **Mountain Map** sheets
- **Foam-Mountain Topographic Map** transparencies

NOTE

The colors used on most USGS topographic maps are:

Green: Plants—forest, brush
Blue: Water—glaciers, lakes, ocean stream
Brown: Land features—lava flow, sand, contour lines
White: No tall plants—desert, prairie
Black: Human-made structures, places

NOTE

Students might find answers to their questions in the reading "Topographic Maps" in the *FOSS Science Resources* book.

INVESTIGATION 3

PART 1 Using Models and Topographic Maps

SESSION 2

17 Practice reading the slope

In groups, have two students at a time each point to a place on the USGS topographic map. The two students not pointing should state which place they think is steeper. Pointers agree or disagree, and discussion continues until all have come to consensus. Then they switch roles. Continue the activity until you feel confident that students can determine relative steepness based on the contour line spacing.

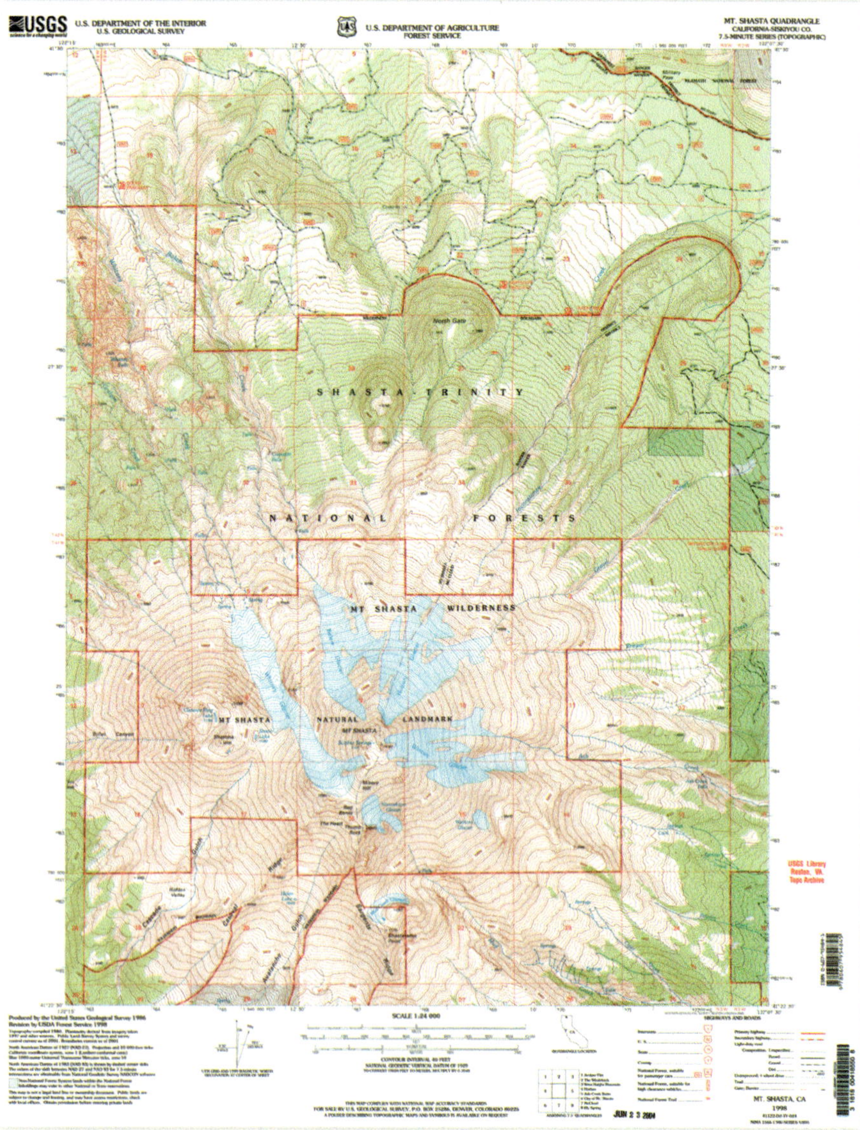

Science and Engineering Practices

- Developing and using models

Crosscutting Concepts

- Scale, proportion, and quantity

NOTE

Explain to students that the simple foam mountain model of Mount Shasta was designed by using this more detailed topographic map. If you have a scale model that incorporates good information about a landform, you can make a map. And you can make a physical model from a topographic map.

18 Have a sense-making discussion

Here is a sense-making discussion template that can be used to facilitate a student-centered discussion. To support inclusion, have students make a circle so they can all see each other. Encourage students to add on to the ideas of others, to state whether they agree or disagree with what other students say, and ask each other to explain their thinking and to provide evidence to support ideas.

NOTE

See the Sense-Making Discussions for Three-Dimensional Learning chapter for more strategies and structures to support productive whole-group discussions.

NEXT GENERATION SCIENCE STANDARDS		
Focus DCI(s)	**Focus SEPS(s)**	**Focus CCs**
Plate tectonics and large-scale system interactions (ESS2.B)	**Develop models** to describe phenomena. **Analyzing and interpreting data** to make sense of phenomena.	Scale, proportion, and quantity

Questions and What to Listen for	
Step 18: Have a sense-making discussion	
What to ask	➜ *What is the contour interval for the topographic map you drew? [500 feet]* (Analyzing and interpreting data) ➜ *Is it the same for the USGS topographic map? [No, it is 40 feet.]* (Analyzing and interpreting data; scale proportion, and quantity) ➜ *Why do you think maps use different contour intervals?* (Developing and using models; scale proportion, and quantity) ➜ *Which side of the mountain shown on the map sheet is steeper? [B] How could you tell by just looking at the topographic map?* (Developing and using models; scale proportion, and quantity) ➜ *Which side of the mountain would you want to walk up and why?* (Developing and using models; scale proportion, and quantity) ➜ *Why might people involved in outdoor activities use topographic maps to plan a hike to a mountain peak?* (Developing and using models; scale proportion, and quantity) ➜ *Why might scientists use topographic maps?* (Developing and using models; scale proportion, and quantity) ➜ *How can we draw the profile of a mountain from a topographic map?* (Developing and using models; scale proportion, and quantity)
What to listen for	• Depending on change in the landscape and the area covered by the map, different contour intervals are necessary. If the land is relatively flat, a smaller contour interval is more useful. • The closer the contour lines, the steeper the mountain. • People use maps and models to help them understand the land.
Scaffolding questions	➜ *Why is it helpful to know if a mountain is steep?*
Application questions	➜ *What are the strengths and limitations of these models?* ➜ *How do maps help us study the surface of Earth?*

PART 1 Using Models and Topographic Maps

contour interval
contour line
elevation
satellite cone
sea level
topographic map
volcano

ENGLISH LANGUAGE SUPPORT
For students who need help getting started, provide a writing frame such as,

In order to represent different elevations of a mountain, you can use a _____. It shows _____. The way it works is _____.

Science and Engineering Practices

- Developing and using models
- Analyzing and interpreting data
- Constructing explanations

Disciplinary Core Ideas

- ESS2.B: Plate tectonics and large-scale system interactions

Crosscutting Concepts

- Scale, proportion, and quantity

SESSION 2

19 Review vocabulary

Review key vocabulary introduced in this part. If these words are not on the word wall, add them now.

20 Respond to the focus question

Have students respond to the focus question in their notebooks.

➜ *How can we represent the different elevations of a mountain?*

Look for these ideas in their responses:

- Topographic maps represent the shape of the landform.
- Each line represents one elevation, for example, 11,000 feet.
- The interval between contour lines is always the same.
- You can tell the steepness of a slope by how close together the lines are; the closer the lines, the steeper the slope.

SESSION 3 45 MINUTES

21 Assess progress: response sheet

Distribute a copy of *Response Sheet A—Investigation 3* and *Response Sheet B—Investigation 3* to each student. This assessment should be completed during class. Collect the response sheets after class and check students' ability to analyze and interpret a topographic map and construct explanations by drawing a trail for hikers given the interests of the hikers and the terrain of the landform.

WHAT TO LOOK FOR

- *Students start the trail at the picnic area near the highway.*
- *Students analyze and interpret the information on the topographic map to determine the steeper path up the mountain that passes the waterfall as part of the criteria.*
- *Students determine a less-steep path for the way back down to the picnic area.*
- *Students construct an explanation about why they drew the path the way they did, including something about how the scale model of the map indicates where the steeper slopes are with contour lines that are closer together.*

Have students return all materials to the materials station. Foam mountains should be taken apart and bagged.

Interpreting Information

22 Read "Topographic Maps" (optional)

Tell students that this article will help them learn more about how topographic maps are made and the standardized symbols used to communicate information. Before reading, have students do a quick write in their notebooks, jotting down three things they know so far about topographic maps and a question they have about how they are made or how to interpret the symbols.

Give students a few minutes to preview the text by looking at and discussing the diagrams with a reading partner.

Have students read the story independently or in guided reading groups. Tell students to think like a cartographer (a person who makes maps) as they are reading and to jot down in their notebooks or on self-stick notes important information they would need to make a map.

To discuss the reading, have students compare their notes to a reading partners' notes. Have students share the features of a topographic map that they learned from the article. Make a class list on chart paper. The list should include:

- Contour lines
- Index contours
- Symbols and colors for landforms and structures
- Scale
- Elevation

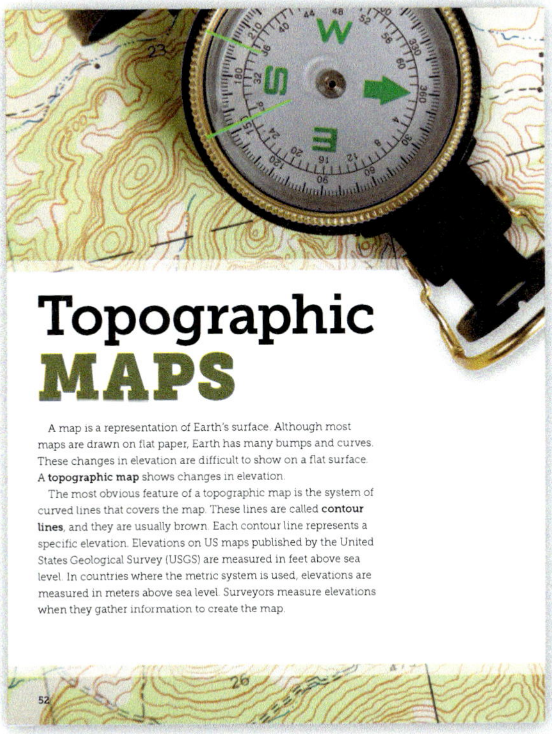

Topographic MAPS

A map is a representation of Earth's surface. Although most maps are drawn on flat paper, Earth has many bumps and curves. These changes in elevation are difficult to show on a flat surface. A **topographic map** shows changes in elevation.

The most obvious feature of a topographic map is the system of curved lines that covers the map. These lines are called **contour lines**, and they are usually brown. Each contour line represents a specific elevation. Elevations on US maps published by the United States Geological Survey (USGS) are measured in feet above sea level. In countries where the metric system is used, elevations are measured in meters above sea level. Surveyors measure elevations when they gather information to create the map.

52

Discuss the Questions for Reflection.

➜ *What information can you get about an area from a topographic map?* [Each contour line represents a specific elevation. Colors show what is on the land (vegetation, water, structures).]

➜ *Who might use a topographic map and how might the map be useful?* [Hikers, ecologists, geologists, volcanologists, construction engineers use topo maps to locate their position at any time and know the elevation for keeping records and designing plans.]

ELA CONNECTION

These suggested strategies address these English Language Arts objectives.

Refer to details and examples in a text when explaining what the text says explicitly and when drawing inferences from text.

Determine the main idea of a text and explain how it is supported by key details; summarize the text.

Explain procedures, ideas, or concepts in a scientific text.

Interpret information presented visually, and explain how the information contributes to an understanding of the text.

Gather relevant information from experiences and print, and categorize the information.

Science and Engineering Practices

- Obtaining, evaluating, and communicating information

NOTE

Discuss the role of the USGS. Visit the USGS website and let students view topographic maps from your local region.

PART 1 Using Models and Topographic Maps

ELA CONNECTION

These suggested strategies address these English Language Arts objectives.

Refer to details and examples in a text when explaining what the text says explicitly and when drawing inferences from text.

Determine the meaning of general academic domain-specific words or phrases.

Science and Engineering Practices

- Obtaining, evaluating, and communicating information

SESSION 3

23 Read "A Cascades Volcano: Mount Shasta"

Have students read the article about Mount Shasta to obtain and communicate information about the mountain they have been mapping. Give students a moment to share with a partner what they have learned so far about Mount Shasta and any experiences they have had observing or exploring other mountains.

24 Use a preview/predict strategy

Before reading, have students independently preview the article, directing attention to the map, photographs, captions, and the words in bold. Then ask students to discuss with a partner and write down three things they think they will learn from the article in their notebooks. Next, have partners verbally share their list with another pair of students and as a group, create one list. Call on a reporter from each group to share their list and together generate a class list on chart paper or the board. Read the article aloud or have students read with a partner or silently. Re-read certain sections and discuss the questions or prompts to check for understanding.

A Cascades Volcano: MOUNT SHASTA

The Cascades are a mountain range that stretches from British Columbia, Canada, down to northern California. The Cascades include at least 16 active volcanoes. An active volcano is one that has erupted in the past 10,000 years.

Some of these active volcanoes are well-known and located near big cities. Mount Rainier is located near Seattle, Washington. Its snowcapped peak towers over the city. Mount Baker is also near Seattle. Many people enjoy hiking, camping, and skiing on these mountains, even though they are volcanoes.

A map showing the Cascade Mountains

The Cascades volcanoes were created when a plate of oceanic crust sank and slid beneath the North American Plate (continental crust). This kind of interaction between crustal plates is called **subduction**, and it injects water and melts rock to form magma. The Cascade Mountains are part of a global system of volcanoes called the Ring of Fire. The process of subduction occurs in many places around the edge of the Pacific Plate. This results in frequent volcanic eruptions and earthquakes in this huge system.

Develop Understanding Point out the word *subduction* and have students paraphrase what it means with a partner. Give them a moment to make a simple diagram of how they visualize this process in their notebooks.

56

An aerial view of Mount Shasta

Mount Shasta is located 65 kilometers (km) south of the California–Oregon border. At 4,317 meters (m), it is the second-highest mountain in the Cascade Range. Mount Shasta formed on the remains of an older volcano that collapsed 300,000 to 500,000 years ago. Its collapse caused a huge landslide.

Mount Shasta is an example of a stratovolcano. Stratovolcanoes are volcanoes composed of lava flows, ash, and other material blown out by explosive activity. Shastina, the prominent cone on the north-west flank of Shasta, was active about 9,200 years ago. Shasta's peak was active only a few centuries ago.

Over the past 300,000 years, Mount Shasta has been quiet for long periods of time. Then it has a short period of activity. Mount Shasta's last major eruption was 3,200 years ago.

Although Mount Shasta has not erupted in a very long time, scientists still observe its activity. Volcanic gases vent out of the top of the volcano. This shows that the volcano is still very hot inside. That tells scientists that the volcano is young in geological time. Hot springs (hot water) also come out of the volcano's top. That is another sign that the volcano is filled with magma. The magma heats the water so that it is hot. It may be pleasantly warm like water in a bathtub or so hot that it could burn you.

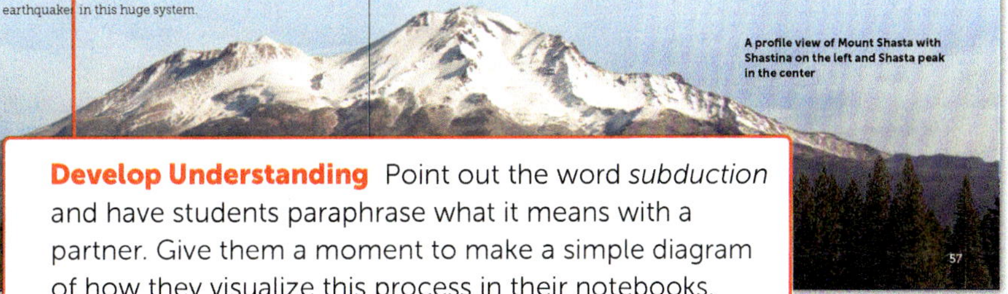
A profile view of Mount Shasta with Shastina on the left and Shasta peak in the center

57

The sides of Mount Shasta are covered with rock and ash. During heavy rain, water can rush down the mountain's side, creating mudflows. Mudflows caused by volcanic eruptions are called **lahars**. Lahars include mud, water, and rocks. Over the past 1,000 years, heavy mudflows and lahars have clogged streams on the mountain.

Another feature of Mount Shasta is the glaciers that cover it with ice year round. In the past few years, those glaciers have gotten much smaller. Scientists think this is because of climate change. The area has had a long period of drought as well. These factors have caused the mountain's Whitney Glacier to shrink by 800 m. That is 25 percent of its original length. Shrinking glaciers mean that the snowpack on top of the mountain is also getting smaller.

Apply Understanding Ask students to share what they think the effects of the shrinking glaciers are on the mountain and the surrounding area.

Scientists have studied Mount Shasta's history and recorded its activity. It does not look like Mount Shasta will erupt in the near future. Earthquakes often occur before a volcano erupts. There has been little earthquake activity at Mount Shasta. That gives scientists another clue that this "active" volcano is quiet for now. But it is still considered an active volcano to be monitored.

Snow and ice cover Mount Shasta in the winter.

Questions for Reflection

1. What caused the formation of the Cascade Mountains?
2. What larger global system are the Cascades part of?

58

SESSION 3

25 Discuss the reading

Refer to the preview/predict class list and discuss what students learned from the reading. Discuss these questions (including the Questions for Reflection) and encourage students to refer to details and examples in the text when answering.

→ *What caused the formation of the Cascade Mountains?* [Interaction between crustal plates. This is called subduction.]

→ *What larger global system are the Cascades part of?* [The Ring of Fire]

→ *Why is Mount Shasta still considered an "active" volcano?* [Volcanic gases and hot springs come out of the top.]

→ *Why are the glaciers on Mount Shasta getting smaller?* [Climate change is causing the temperature to rise and the amount of snow to decrease on Mount Shasta. This causes the ice to melt so the glaciers are shrinking.]

26 Discuss a southwest route to the Shasta summit

Ask students if they have a solution to the hikers' problem. Have them describe their suggested route to climb to the summit using the topographic map and the poster and give their reasons.

Describe one classic route: the Avalanche Gulch route. Have students find the route on the USGS topographic map as you describe the landmarks of the Avalanche Gulch route.

- The **Avalanche Gulch** route begins at Bunny Flat trailhead and leads to Horse Camp just below the tree line at 7,900 feet. This is off the topographic map, but Avalanche Gulch is easy to find at the bottom of the map.
- Then the route starts to get steep as you continue to a location called **Helen Lake** at 10,443 feet. The lake is covered in snow during most of the hiking season.
- From Helen Lake, the route climbs up and to the right of **The Heart**, over the **Red Banks** on the right side.
- The next climb is over **Misery Hill**. It starts at over 13,000 feet.
- Then hike across a summit plateau to the summit **peak** at 14,162 feet.

Although this is a non-technical route, it does require some climbing gear (crampons, mountain axe, helmet) and the skills of traveling over snow and ice.

Show the Avalanche Gulch photos that are on The Mount Shasta Avalanche Center website. Those photos provide the view from Bunny Flat, the view from I-5, late Spring/early Summer, and late summer season (not a good time to climb) will show students the volcanic mountain features throughout the year.

NOTE

Route information, including maps and photos of the Avalanche Gulch as well as descriptions of other routes, is available on the Mount Shasta Avalanche Center website. The center functions as a partnership between the US Forest Service and the not-for-profit organization, Friends of the Mount Shasta Avalanche Center.

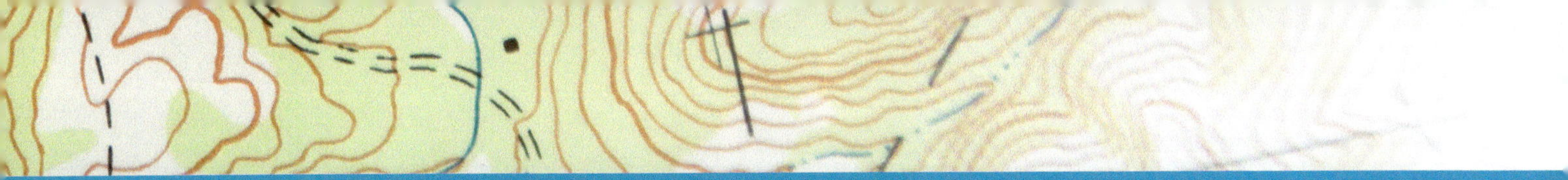

Reflecting on a Phenomenon

27 Describe what map the hikers should use

This a good time to have students add to their thinking about how information about landforms like Mount Shasta is represented in the mountain photograph, the physical model, and the topographic map. Have students add to their explanation about how the topographic map will help the hikers plan their route to the summit. Review the driving question.

→ *What kind of a map do the hikers need and how will it help them to plan a route to the mountain peak?*

Students should say that the hikers will get the most detailed information from a topographic map to plan and guide the hike. Here are the characteristics of the different maps students can compare to support their explanations.

- The photograph provides a visual description with overall shape but does not provide specifics about elevation or slope.
- The scale model is three-dimensional and provides shape and details general elevation with contour intervals but is limited in the amount of detail.
- The topographic map is flat, two-dimensional, and provides bird's-eye view with detailed contour lines and contour intervals providing information about slope; color provides information about ground cover and vegetation, water, glaciers, creeks, streams, springs structures, roads, trails, and much more. At different times of the year and depending on the weather, snow may cover some of the land.

Students might also add that with these tools, it is important to listen to weather reports and talk with experts who have made the climb recently and can report on the condition of the mountain.

NOTE

See the **Home/School Connection** for Investigation 3 at the end of the Differentiation Opportunities. This is a good time to send it home with students.

28 Discuss response sheet (optional)

Conclude Part 1 by having students discuss their response sheet ideas. Have them explain their routes up the peak and down from the peak and the reasons they selected the route.

Science and Engineering Practices

- Developing and using models
- Analyzing and interpreting data
- Using mathematics and computational thinking
- Constructing explanations

Disciplinary Core Ideas

- ESS2.B: Plate tectonics and large-scale system interactions

Crosscutting Concepts

- Patterns
- Scale, proportion, and quantity

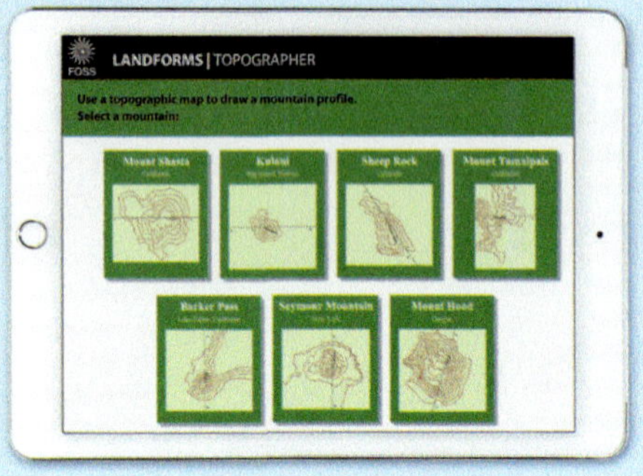

SIDE TRIP 3

Drawing a Profile

Focus on mathematics and scaling

Purpose

This is an optional activity, a side trip from the main pathway, that provides an extended experience in mathematics and scaling that can be used with the whole class, or it can be used for a student choice challenge, or for differentiation.

- The detailed lesson plan is a printable digital resource found on FOSSweb including notebook master 17, *Contours and Intervals*, teacher master 7, *Mountain Map with Elevations*, and teacher master 8, *Profile*.
- The online activity, "Topographer," is on FOSSweb.

FOCUS QUESTION

How can we draw the profile of a mountain from a topographic map?

Activity Summary

Students use their topographic maps to produce two-dimensional profiles, or cross-sections, of their foam mountain. The profile reveals a side view of Mount Shasta, a dormant volcano.

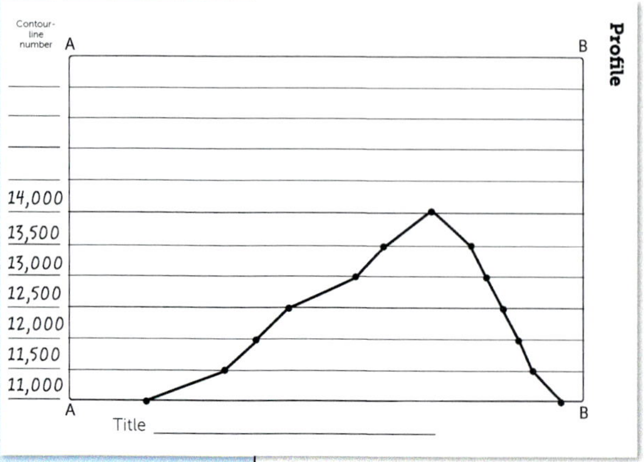

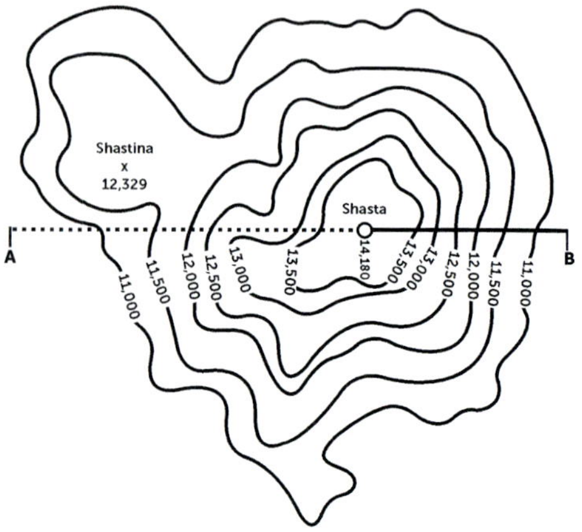

PART 2 Mount St. Helens Case Study

Teacher Background

Science and Engineering Practices

- Developing and using models
- Analyzing and interpreting data
- Constructing explanations
- Engaging in argument from evidence
- Obtaining, evaluating, and communicating information

Disciplinary Core Ideas

- ESS2.B: Plate tectonics and large-scale system interactions
- ESS3.B: Natural hazards

Crosscutting Concepts

- Scale, proportion, and quantity
- Stability and change

Phenomenon Storyline

LEARNING OBJECTIVES:

- **Students ask questions and use maps and scale models to gather data about mountain landforms using scale, proportion, and quantity.**
- **Students collect, analyze, and interpret data to describe stability and change to a mountain after a volcanic eruption.**

Part 1 begins the use of representational tools to describe the structure of a volcanic mountain. In Part 2, students apply these tools to make a claim and support it with evidence. The scenario is: Students were looking at maps and images of the same volcanic mountain at different times. They observed some major differences in the landforms. The driving question (and the focus question) for this investigation is

How did the mountain change and what caused the change?

Here are some things students will find out to connect to the storyline.

- Maps and photographs developed at different times are useful in providing evidence that earth material has moved and has caused changes in a landform over time.
- Mountains that were formed through volcanic activity and are active need to be monitored.
- An event such as a volcano eruption provides the opportunity to measure change in a mountain due to a natural event.

FOCUS QUESTION

How Did the Mountain Change and What Caused the Change?

Along the west coast of the United States—in California, Oregon, and Washington—volcanoes pose a threat of rapid change to Earth's surface. Indeed, Mount St. Helens changed the shape of 230 square miles in a matter of 10 minutes on May 18, 1980. Volcanoes are the result of a tremendous buildup of heat and pressure in the mantle below Earth's crust. Eventually the pressure forces the melted rock called magma upward through a crack or weak spot in the crust. When it does, the magma pours out as lava or pyroclastic flows (mixtures of extremely hot gases and dry rock fragments). This material and volcanic ash flow down the side of the volcano or shoot up into the air to be carried by wind currents before being deposited some distance from the eruption. Volcanoes destroy the landforms at the site of the eruption, but they create new landforms through deposition of new rock material on Earth's surface. Eruptions are constructive events because the earth material they leave behind is not simply relocated surface material, but completely new material. The result of a volcanic eruption is the creation of brand-new landforms in a very short period of time and over the long run volcanic deposits also benefit soil, creating some of the richest soil on Earth for growing food crops.

The conceptual flow for Part 2 starts with students making observations of Mount St. Helens, an active volcano. By comparing models created at different times in the history of the mountain, students observe the changes that took place in recent times as a result of the volcanic eruption. Students use topographic maps of the mountain before and after the 1980 eruption to determine the effect of this natural hazard event.

What representation of a mountain can help hikers plan a route to the summit?

PART 1

Earth materials form natural structures called landforms.

Mountains are one kind of landform.

A topographic map is a visual display of Earth's surface.

Some mountains are volcanoes.

Contour lines indicate elevation.

How did the mountain change and what caused the change?

PART 2

Some events change Earth's surface very quickly.

Heat and pressure in the Earth's mantle below the crust cause eruptions.

Eruptions are natural hazards.

INVESTIGATION 3

PART 2 Mount St. Helens Case Study

Materials

PROVIDED EQUIPMENT

For each student

1 *FOSS Science Resources: Soils, Rocks, and Landforms*
"What Geoscientists Do" (optional)

For the class

1 Poster, *Mount St. Helens*

FOSSWEB RESOURCES

For each student

- Teacher master 9, *Topographic Map A*
- Teacher master 10, *Topographic Map B*

For the class

- Word wall cards and equipment cards
- Teaching slides
- Anchor Phenomenon Resource Video 2

For the teacher

- *Performance Assessment Checklist*

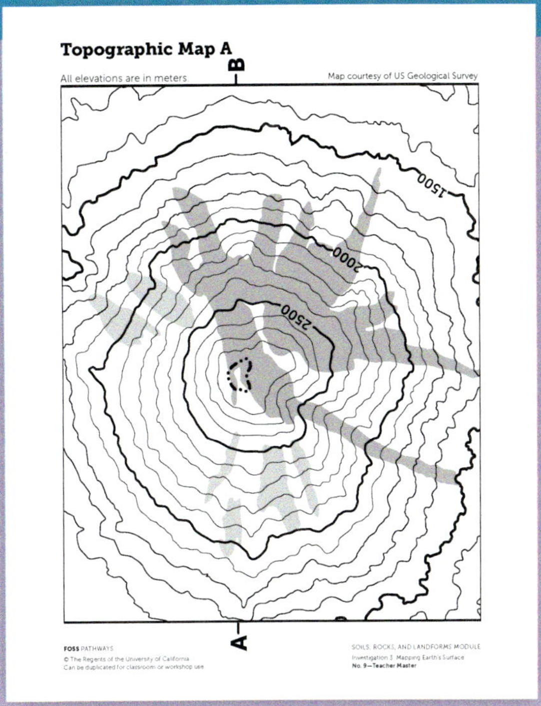

No. 9—Teacher Master

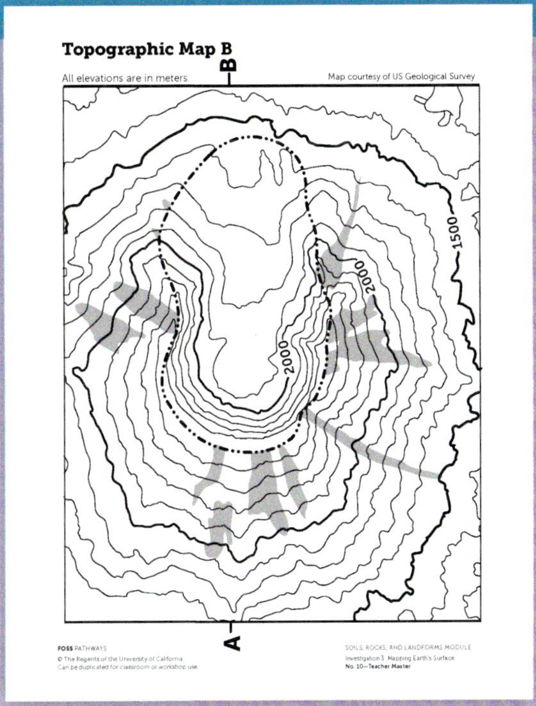

No. 10—Teacher Master

NOTE

These topographic maps are scaled in meters. For comparison, Mt. Shasta is 4,332 m high.

INVESTIGATION 3

PART 2 Mount St. Helens Case Study

Getting Ready

Schedule

1 Session
Session 1: Steps 1–14 (Steps 9, 11, and 12 optional)

Review

- Background for Part 2
- FOSSweb Resources for this part
- Reading: "What Geoscientists Do" (optional)
- Video: *Mount St. Helens' Impact* (USGS video)
- Teacher Preparation Video
- Anchor Phenomenon Resource Video 2

Print or Copy

For each student:

- Teacher master 9, *Topographic Map A*
- Teacher master 10, *Topographic Map B*

For the teacher:

- *Performance Assessment Checklist*

Plan for Assessment

- Review Step 2, "What to Look For" in the performance assessment. Students will use topographic maps to look at the changes that occurred in the 1980 eruption of Mount St. Helens. This will provide opportunities for you to observe students' science practices.

INVESTIGATION 3

PART 2 Mount St. Helens Case Study

Guiding the Investigation

SESSION 1 **45 MINUTES**

1 Introduce new topographic maps

Distribute one copy of teacher masters 9 and 10, *Topographic Map A* and *Topographic Map B,* to each group. Ask students to study the two maps (models of a system) and then engage in argument in their groups to determine if the two maps are the same mountain or two different mountains.

2 Assess progress: performance assessment

Circulate from group to group. Listen to the group discussions and observe how students argue whether the maps describe the same mountain. Write notes on the *Performance Assessment Checklist* while students work.

WHAT TO LOOK FOR

- *Students argue for or against the supposition that the maps are describing the same mountain; they use evidence from analyzing the topographic maps to support their claims.*

3 Introduce the anchor phenomenon—impact to Mount St. Helens

Have groups report their conclusions to the class, including their reasons for those conclusions. If students don't come to the conclusion that these could be the same mountain, point out some of the prominent patterns on the south, east, and west sides that are almost the same on both maps. It would be very unusual for different mountains to have features that similar. Then acknowledge that the north side of the mountain on map A is very different from the north side of the mountain on map B.

Confirm that both of these maps are of Mount St. Helens, a volcano that is part of the same mountain range as Mount Shasta. They are both part of the Cascade mountain range on the West Coast of the United States. Other students also determined these maps were of the same volcanic mountain showing the mountain at different times. They, like you, observed some major differences in the landforms.

Mount St. Helens erupted on May 18, 1980. Before then, the last time Mount St. Helens had erupted was 1857. It had been stable and quiet for more than 120 years.

Science and Engineering Practices

- Developing and using models
- Analyzing and interpreting data
- Engaging in argument from evidence

Disciplinary Core Ideas

- ESS2.B: Plate tectonics and large-scale system interactions

Crosscutting Concepts

- Scale, proportion, and quantity
- Stability and change

ENGLISH LANGUAGE SUPPORT

Use the map from the Mount Shasta article to show students where Mount St. Helens is located.

NOTE

Use the Anchor Phenomenon Resource Video 2 for this introduction.

4 Introduce the focus/driving question

FOCUS QUESTION

How did the mountain change and what caused the change?

Ask the focus question (also the driving question) as you project or write it on the board. Have students write the date and focus question in their notebooks.

➜ *How did the mountain change and what caused the change?*

NOTE

The focus question and the driving question are the same for this part.

5 View the USGS video of the eruption

Tell students,

On March 20, 1980, a moderate (4.2) earthquake shook Mount St. Helens. Scientists took note, but became even more curious about the mountain when they noticed steam vents a week later. By the end of April, the mountain began to bulge. Scientists were using a new, supposedly very precise instrument to measure changes in the mountain, but they just couldn't believe that the bulge was growing about 2 meters a day. These measurements were so out of the ordinary that they sent the instrument back to the factory to get fixed. It turned out the instrument was working perfectly. Magma was building up inside the mountain at a very rapid rate. The eruption of Mount St. Helens was an event that scientists anticipated, but completely underestimated. Let's watch a video in which USGS scientists describe what happened on May 18, 1980.

Show students the *Mount St. Helens' Impact* video on FOSSweb to obtain, evaluate, and communicate information about this phenomenon.

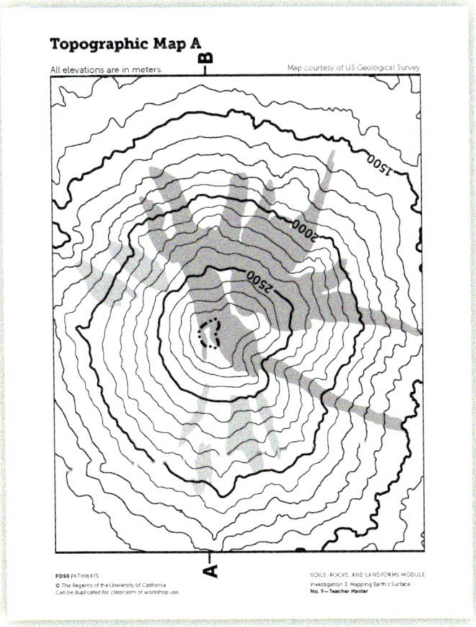

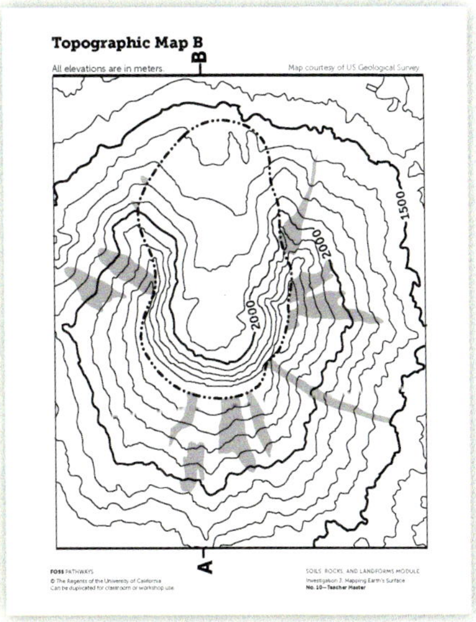

ENGLISH LANGUAGE SUPPORT

As you narrate, make simple labeled drawings of the events to assist with comprehension.

Science and Engineering Practices

- Obtaining, evaluating, and communicating information

NOTE

The AB line runs North to South.

SESSION 1

6 Discuss the video

Use these questions as needed to guide the discussion. Included are a few additional facts that you can use to emphasize how unusual and awesome this event was.

→ *What did you find most interesting about the video?*

→ *When most people think about volcanoes, they think about liquid lava splashing out of the top of a mountain with red and black lava flowing slowly down its sides. How was the eruption of Mount St. Helens different?* [Instead of flowing lava, the magma in Mount St. Helens exploded in a pyroclastic flow that flattened everything in its path (some 600 square kilometers) in less than 10 minutes; huge trees were knocked over as if they were toothpicks.]

→ *Several people who lived near Mount St. Helens at the time refused to leave when scientists warned them to go. Why do you think they decided to stay?* [Mount St. Helens had been inactive for so long they didn't believe what the scientists were telling them. They didn't think that it would really erupt.]

→ *Why were scientists so eager to get out on the mountain after each eruption after the big one in May 1980?* [The eruptions provided the rare opportunity to observe a volcano when it was active; they provided the opportunity to learn how to reduce the impacts that volcanoes might have on people. Scientists want to be able to warn people of future eruptions so they can get out of the way.]

Have them discuss how the mountain changed. Have a few students share their groups' observations with the class. Students should say:

- An eruption occurred that changed the mountain. A big part of the mountain is gone. It broke and eroded down away from the mountain.
- Supporting data is shown on the before and after topographic map.

You can offer this information.

Volcanic eruptions are generally thought of as a mountain-building process. One of the things that scientists did not expect when Mount St. Helens erupted was blowing off its top, reducing its height by about 400 meters, nor did they expect the collapse of the north side, which caused the mudflows and a great deal of damage to trees and human structures such as bridges, highways, and homes.

7 Discuss volcanic eruptions as natural hazards

Natural hazards are events in the natural world/environment that are harmful to people, like a volcanic eruption. The eruption of Mount St. Helens had a large economic impact. Everything within

NOTE

Keep this video at the ready to show again in Step 9.

NOTE

Pyroclastic flows consist of rocks and extremely hot gases that flow downhill, usually hugging the ground. In the case of Mount St. Helens, it included ash catapulted into the atmosphere that traveled for kilometers before settling back to Earth. The word "pyroclast" comes from two Greek words meaning "fire" and "broken pieces."

Crosscutting Concepts

- Stability and change

that 600-square-kilometer area on the north side was destroyed, including lumber camps and more than 200 homes. In addition, 57 human lives were lost.

Students can look at pages 70–71 of the article "Events That Change Landforms" in *FOSS Science Resources* to find out more.

8 Have a sense-making discussion

Have students discuss their initial responses to the focus question. You can use these questions to guide the discussion.

- → *What did Mount St. Helens look like before the eruption? After?* [Before it looked pretty much the same from every angle. Then a bulge appeared on the north side and there was a lateral blast with rock, ash, and gases exploding out horizontally from the north side. The topographic maps before and after showed the change in the landform on the north side. It left a huge depression or hole on the north side. The pyroclastic flow killed all life in its path.]
- → *What happened to the earth material that was inside the volcano?* [The earth materials exploded out as rock, ash, and gas and became a pyroclastic flow.]
- → *How long did the changes take to occur?* [The eruption lasted about 9 hours but the bulge appeared 2 months earlier.]
- → *What did scientists do before and after the eruption?* [During and after, scientists monitored for earthquakes and took videos and pictures of the bulge on the north side and gases that came from the volcano. After, they forecasted eruptions and raced to study hot, fresh deposits from the explosive eruptions. Helicopters carried them from place to place and kept them safe.]
- → *Why is it important for scientists to monitor changes in volcanic mountains?* [The equipment used to observe and monitor changes in volcanoes can make predictions about future volcanic activity to reduce the impact of eruptions.]

9 Watch the USGS video one more time (optional)

Have students keep their notebooks open to take notes as you have the class watch the *Mount St. Helens' Impact* video one more time. This time, have students listen for things that scientists were doing at the time and have done since then to obtain information about this event.

Have students talk with a partner and compare the notes they took when they watched the video a second time. Ask them to discuss why the eruption of Mount St. Helens was such an important event for geoscientists and engineers.

10 Respond to the focus/driving question

Have students respond to the focus/driving question in their notebooks.

→ *How did the mountain change and what caused the change?*

Crosscutting Concepts

- Stability and change

ENGLISH LANGUAGE SUPPORT

Review language structures for cause and effect such as,

When, _____ then, _____. The change was caused by _____.

For students who need help getting started, provide sentence frames such as,

Mount St. Helens was changed by _____. First, _____. Then, _____. As a result, _____. It's important to _____ because _____.

NOTE

If students are interested in finding out more about volcanoes, introduce them to the USGS and NOAA websites with maps and current information about volcano activity alerts. Refer to the environmental extensions in the Differentiation Opportunities at the end of the investigation.

INVESTIGATION 3

PART 2 Mount St. Helens Case Study

Interpreting Information

ELA CONNECTION

These suggested strategies address these English Language Arts objectives.

Determine the meaning of general academic domain-specific words or phrases.

Interpret information presented visually, and explain how the information contributes to an understanding of the text.

Gather relevant information from experiences and print, and categorize the information.

Determine or clarify the meaning of unknown and multiple-meaning words and phrases.

Science and Engineering Practices

- Obtaining, evaluating, and communicating information

What Geoscientists Do

Where do **geoscientists** go to do their work? Just about anywhere. That's because the prefix *geo-* means Earth. Geoscientists observe, investigate, and test landforms all over Earth. They work to discover, manage, and protect Earth's natural systems. Earth's rocks, minerals, soils, water, air, plants, animals, and fossil fuels together are called **natural resources**. Geoscientists study the history, distribution, use, and conservation of Earth's nonliving natural resources.

Most geoscientists spend a lot of time doing fieldwork. That means they are outside in direct contact with Earth. Most geoscientists have a specialty. They focus on one part of the Earth system. They use instruments to collect data about the Earth system they study.

Visit FOSSweb on ThinkLink to explore the Science and Engineering Career Database.

Geoscientists work outside studying Earth's natural systems and natural resources.

59

SESSION 1

11 Read "What Geoscientists Do" (optional)

Explain to students that the article, "What Geoscientists Do," describes the various careers followed by people who enjoy studying Earth and earth materials. It also describes the ways that engineers contribute to the work of geoscientists.

Have students read independently or with a partner to obtain and communicate information about geoscientists. Provide self-stick notes for students to jot down unknown words or phrases and questions they have about the different types of geoscientists.

Discuss any of the unknown words and phrases. Ask students to discuss how they figured out some of the words using context clues, root words, and affixes.

12 Engage in pair conversations (optional)

After reading, have students discuss the reading in pairs. Students should select the career that interests them the most, and describe it to their partners. If possible, students should include the kinds of instruments that the scientists use and their collaborations with engineers. They should speculate on what they might need to learn to be good at that career.

The career descriptions include:

- Seismologist
- Marine geologist
- Atmospheric scientist
- Structural geologist
- Hydrologist
- Environmental geologist
- Volcanologist
- Soil scientist

Have students look again at the "Soil Scientists" and "Soil Science at Home" sections on pages 63 and 64. Discuss how this information relates to the soil explorations students did in the previous investigations. Students might also share their experience with composting.

Reflecting on a Phenomenon

13 Finalize the anchor phenomenon response

Have students revisit their explanation of the anchor phenomenon (the driving question and focus question are the same).

→ *How did the mountain change and what caused the change?*

Students should describe the cause of the eruption—the buildup of pressure and heat in the Earth's mantle pushing the melted rock (magma) upward through cracks in the Earth's crust. An earthquake shook the mountain and the bulge that had formed in the north side of the mountain erupted, sending hot gases and rock down the side of the volcano in a mudslide and ash high into the air. When it was over, the north side of the mountain was gone as earth material had been pushed out. The mountain was reduced in height by 400 meters. An area over 600 square kilometers had been flattened and there was much damage to the forest and human structures.

14 Review Investigation 3

Distribute one or two self-stick notes to each student. Ask students to cut each note into three pieces. Have them take 3 minutes to look back through their notebook entries to find the three most important things they learned in Investigation 3. They should tag those pages using the three self-stick notes.

When students have completed their notebook review and placed their tags, lead a short class discussion to create a list of three-dimensional statements that summarize what students have learned in this investigation.

- We used topographic maps to model the shape of the land; each contour line shows the shape of the land at a certain elevation; the closer the lines, the steeper the slope.
- We obtained and evaluated information provided by topographic maps about a landform to plan a hike to a mountain summit.
- We obtained and evaluated information provided by topographic maps to communicate the impact of a volcanic eruption.
- We obtained information about how scientists and engineers study changes in landforms in order to make predictions about future natural events that might cause damage.

Science and Engineering Practices

- Constructing explanations

THREE-DIMENSIONAL LEARNING OBJECTIVES

- **Students ask questions and use maps and scale models to gather data about mountain landforms using scale, proportion, and quantity.** (Parts 1–2)
- **Students collect, analyze, and interpret data to describe stability and change to a mountain after a volcanic eruption.** (Part 2)

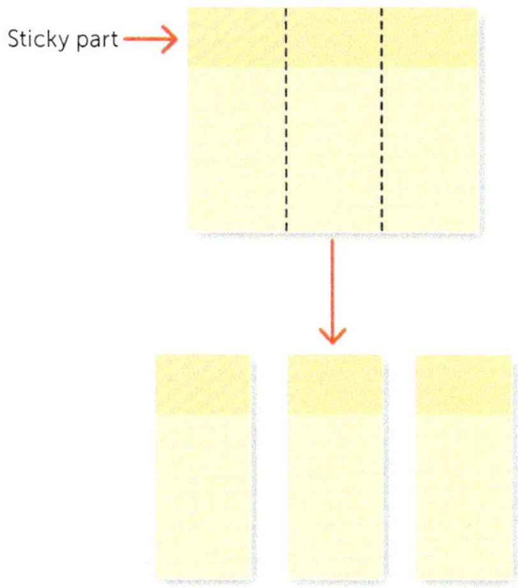

Differentiation Opportunities

Below is a sample of student-choice activities related to this investigation for group or individual projects. Students can select an activity that is of special interest to them, or the class can choose a project for collaboration. Extensions and project ideas are written for students and accessible on FOSSweb.

Language Extensions

- Interview a surveyor
- Write a fictional story, and create a treasure map

Science Extensions

- Make a profile map of Mount St. Helens before and after the eruption
- Use the online activity "Topographer" on FOSSweb

Engineering Extension

- Design a tool to measure different slopes around your schoolyard

Environmental Extensions

- Hike or bike a local mountain trail using a topographic map
- Review the USGS Volcano Hazards website that provides maps and current information about volcano activity

Home/School Connection

Students apply what they have learned to interpret a topographic map that describes the FOSS Creek area. Print or make copies of *Home/School Connection* for Investigation 3 and send it home with students after Part 1.

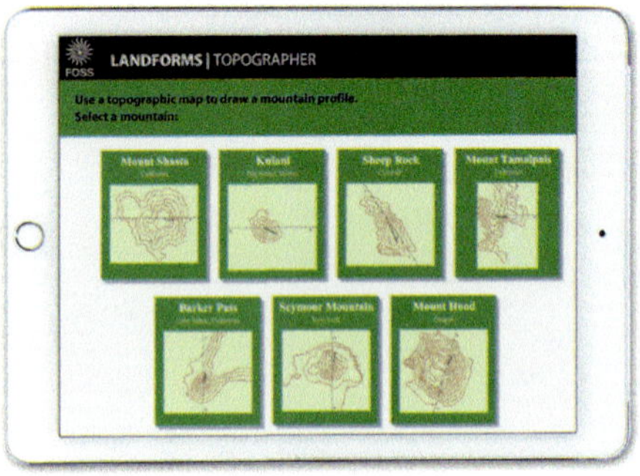

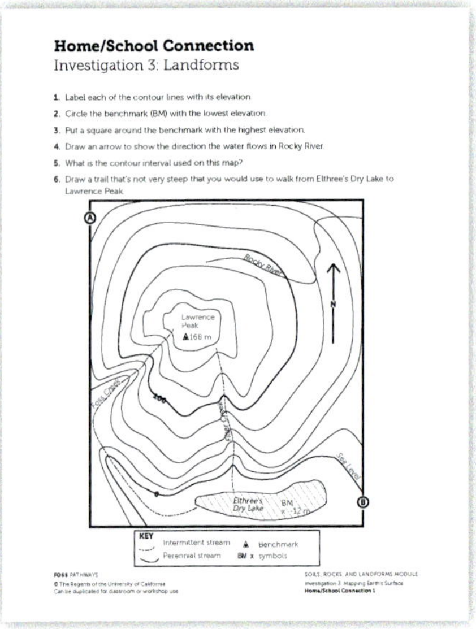

Home/School Connection
Investigation 3: Landforms

1. Label each of the contour lines with its elevation.
2. Circle the benchmark (BM) with the lowest elevation.
3. Put a square around the benchmark with the highest elevation.
4. Draw an arrow to show the direction the water flows in Rocky River.
5. What is the contour interval used on this map?
6. Draw a trail that's not very steep that you would use to walk from Elthree's Dry Lake to Lawrence Peak.

FOSS PATHWAYS
© The Regents of the University of California
Can be duplicated for classroom or workshop use
SOILS, ROCKS, AND LANDFORMS MODULE
Investigation 3: Mapping Earth's Surface
Home/School Connection 1

No. 3—Home/School Connection

INVESTIGATION 4

Changes to Landforms

Investigation 4

INVESTIGATION 4

Changes to Landforms

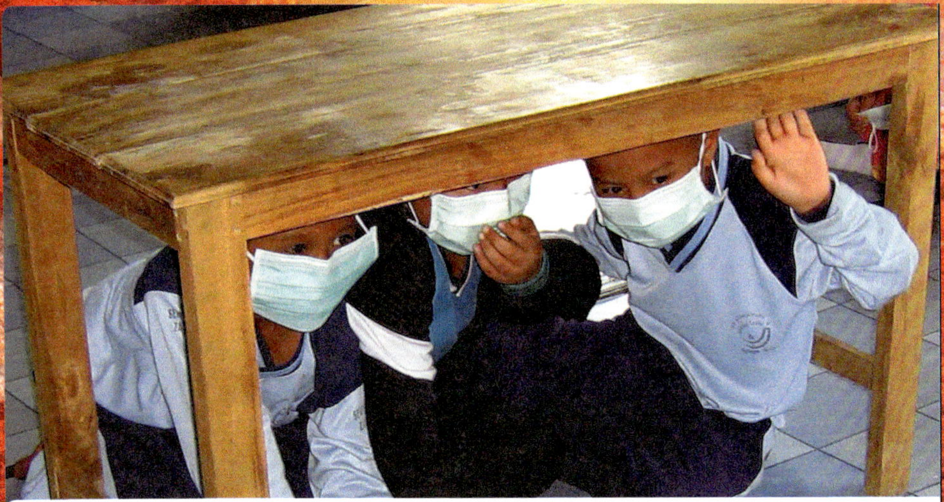

Phenomenon Storyline

Earth-shaking events, Part 1: Students in a classroom in South-Central Alaska dive under their desks when the building and its contents start shaking. They wait until the vibrations stop and evacuate with their teacher to the outdoors. People in the northern part of the state didn't feel the shaking.

Where do earthquakes and volcanoes occur around the world?

Rock layers, flat and tilted, Part 2: A family visits a natural outdoor site and finds large rock-layer formations with unusual stripes of red, pink, and yellow. Some of the layers are flat and others are tilted. In another site they discover fossils.

What do rock layers and fossils tell us about the history of a place?

Students have used mapping tools to find out about volcanic mountains that are part of the Cascade Range. Through a case study, they gathered information about the changes that took place when Mount St. Helens erupted.

In this investigation, students use mapping tools and data sets to reveal patterns to determine where landform-changing events occur on Earth. Students interpret rock layers and integrate the slow processes of weathering, erosion, deposition, canyon-forming (downcutting), and uplifting, which result in the rock formations we observe on the surface of Earth today.

THREE-DIMENSIONAL LEARNING OBJECTIVES

- Students analyze and interpret data sets of active volcanoes and large earthquakes by mapping locations on a map to reveal patterns. **(Part 1)**
- Students obtain information from media to construct explanations about the geological history revealed by patterns of rock layers and the presence of types of fossils. **(Part 2)**

INVESTIGATION 4

Changes to Landforms

At a Glance

Investigation Part	Sessions	Practices and Crosscutting Concepts
Introduce anchor phenomenon 3—Earth-shaking events **PART 1 Mapping Earthquakes and Volcanoes** **FOCUS QUESTION: Where do earthquakes and volcanoes occur around the world?** Students use latitude and longitude coordinates to map the locations of 20 active volcanoes and 20 large earthquakes around the world. They analyze the data and observe that many of these events occur in a pattern on the edges of the Pacific Ocean. Students use the online USGS Latest Earthquakes Map displaying the Earth's tectonic plate outlines to gather information about the cause of these effects. Students view a content video to observe, evaluate, and communicate information about volcanoes and earthquakes.	**SESSION 1 *** Steps 1–10 **SESSION 2** Steps 11–21 (Steps 19, 22 optional) *A session is 45 minutes.	**Practices** • Asking questions • Developing and using models • Analyzing and interpreting data • Constructing explanations • Obtaining, evaluating, and communicating information **Crosscutting Concepts** • Patterns
Introduce anchor phenomenon 4—Rock layers, flat and tilted **PART 2 Interpreting Rock Layers** **FOCUS QUESTION: What do rock layers and fossils tell us about the history of a place?** Students gather information through case studies of phenomenal sedimentary rock formations in a desert. In one case study, slanted rock in sandstone layers expose the effect of tremendous earthquakes or tectonic forces that can bend, twist, break, and uplift rock layers to expose them in an otherwise flat area. A second case study reveals trilobite fossils in sedimentary shale providing evidence that this area was once a marine coastal environment. The mystery is how the shale ended up forming a mountain in the middle of a desert. Several rock-layer puzzles provide an opportunity for students to construct explanations.	**SESSION 1** Steps 1–5 **SESSION 2** Steps 6–7 **SESSION 3** Steps 8–11 **SESSION 4** Steps 12–18 **SESSION 5** Step 19, *Posttest*	**Practices** • Asking questions • Developing and using models • Constructing explanations • Obtaining, evaluating, and communicating information **Crosscutting Concepts** • Patterns • Stability and change

Performance Expectations

4-ESS1-1: Identify evidence from patterns in rock formations and fossils in rock layers to support an explanation for changes in a landscape over time.

4-ESS2-2: Analyze and interpret data from maps to describe patterns of Earth's features.

Disciplinary Core Ideas	Interpreting Information	Assessment
ESS2.B: Plate tectonics and large-scale system interactions **ESS3.B:** Natural hazards • Most earthquakes and volcanoes occur in patterns near the edges of continents. • A world map with latitude and longitude lines is a tool to arrange data of earth-changing events to look for patterns. • The Pacific Ring of Fire contains about 75 percent of the world's active volcanoes and is the source of about 90 percent of the world's earthquakes (USGS). • The surface of Earth is constantly changing; sometimes those changes take a long time to occur and sometimes they happen rapidly. • Catastrophic events have the potential to change Earth's surface quickly.	**Science Resource Book** "Events that Change Landforms" (optional) **Other Media** Online activity: "Latest Earthquakes" Online poster: "This Dynamic Planet" Video: *Earthquakes and Volcanoes* Video: *Shake Alert: Earthquake Early Warning Systems for the West Coast of the United States* Video: *Remembering Ridgecrest* (optional) Anchor Phenomenon Resource Video 3	**Embedded Assessment** Science notebook entry **NGSS Performance Expectation addressed** 4-ESS2-2
ESS1.C: History of planet Earth • Fossils provide evidence of organisms that lived long ago as well as clues to changes in the landscape and past environments. • Sometimes sedimentary rocks contain fossils. • Sedimentary rocks form in horizontal layers and those layers provide information about the geological history of an area. Earthquakes can tilt those layers. • Landforms are constantly changing due to weathering, erosion, deposition, downcutting, and uplifting.	**Science Resource Book** "Valley of Fire" "Fossils Tell a Story" "Frenchman Mountain" **Other Media** Online activity: "Rock-Layer Movie Maker" Anchor Phenomenon Resource Video 4	**Embedded Assessment** Science notebook entry **Benchmark Assessment** *Posttest* **NGSS Performance Expectation addressed** 4-ESS1-1

PART 1 Mapping Earthquakes and Volcanoes

Science and Engineering Practices

- Asking questions
- Developing and using models
- Analyzing and interpreting data
- Constructing explanations
- Obtaining, evaluating, and communicating information

Disciplinary Core Ideas

- ESS2.B: Plate tectonics and large-scale system interactions
- ESS3.B: Natural hazards

Crosscutting Concepts

- Patterns

Aa

coordinates	**longitude**
degrees	**subduction**
epicenter	**tectonic plate**
latitude	

Teacher Background

Phenomenon Storyline

LEARNING OBJECTIVE:

- **Students analyze and interpret data sets of active volcanoes and large earthquakes by mapping locations on a map to reveal patterns.**

Students and their teacher in a classroom in Alaska dive under their desks when the building starts shaking. They wait until the vibrations stop and quickly evacuate the school. The driving question is:

Where do earthquakes and volcanoes occur around the world?

Here are some things students will find out to connect to the storyline.

- By plotting the locations of active volcanoes and strong earthquakes on a world map, patterns are revealed.
- From the data, it seems that earthquakes and volcanoes occur in patterns near the edges of continents.

FOCUS QUESTION

Where Do Earthquakes and Volcanoes Occur around the World?

Volcanoes and earthquakes create violent tremors. A cause of both events is the movement of tectonic plates in Earth's crust. When these plates move, they collide, move apart, or slide past each other. The movement of these plates releases energy in the form of waves that cause rock slabs to break and the ground to shake. The boundaries between two plates is a fault line. Plate movements can also produce conditions that cause volcanoes to form.

Earthquakes are described by their epicenter and depth. The epicenter is the surface point above which the underground rock in the crust breaks. The depth of the break ranges from Earth's surface to about 800 kilometers (km) deep. The deeper the source of the earthquake, the less the effect it has on Earth's surface. The depths of earthquakes provide important information about Earth's structure and the tectonic plates where earthquakes occur.

Human activity can cause earthquakes. Water stored in reservoirs, surface and underground mining, and fracking—the process of injecting liquid at high pressure into underground formations to extract oil or gas—have been shown to produce earthquakes.

Magnitude is the size of the earthquake. Earthquakes below a magnitude of 3 can be felt by humans if they are shallow or near the epicenter. The largest earthquake in recorded history was in 1960 near Valdivia, Chile. USGS recorded its magnitude at 9.5.

Most volcanoes also form along the boundaries of Earth's tectonic plates. When tectonic plates collide, one plate may go below the other. This is called subduction. As one landmass dives deep below the other landmass, the temperature and pressure increase, causing the rock to turn into magma that can seep up to Earth's surface. One thing all volcanic areas have in common is a magma source that delivers molten material and heat to the surface. Interactions between plates along their boundaries provide the materials and energy to shake the earth and create volcanic activity. Most continental and island-arc volcanoes occur along or near where one plate is being pulled, or subducted, into the mantle. These include Mount St. Helens in Washington State and Mount Pinatubo in the Philippines. Another way volcanoes can form is through hotspots that are in the interior of a tectonic plate. This may be how the chain of Hawaiian islands formed.

The area that has produced about 75 percent of Earth's active volcanoes and 90 percent of Earth's earthquakes is around the edges of the Pacific Plate in the Pacific Ocean. It is here that the Pacific Plate interacts with a number of other plates. This horseshoe-shaped area (the Ring of Fire) starts in the southern part of the South American continent, goes north along the western edge to the North American continent, goes east across the Bering Strait then south to Japan, and along the many islands off the Asian continent south to New Zealand. The Cascade Range and the Alaska Aleutian Islands chain are part of the Ring of Fire. Although the volcanoes are caused by the movement of the plates, they are not geologically connected. Thus, if there is an eruption in one of these volcanoes, that doesn't mean that others will erupt.

According to USGS, there are about 1,350 potentially active volcanoes worldwide with 161 of them in the United States. An active volcano is one which is either erupting or is likely to erupt in the future.

Dangers from earthquakes include landslides, mudslides, rockfalls, cracks in the surface of Earth, settling, and the collapsing of structures. Earthquakes can also cause tsunamis, a large displacement of water that causes waves that travel long distances and can release large amounts of energy when the waves reach land. Volcanoes cause tsunamis too.

Volcanoes cause many of the same natural hazards as earthquakes. They harm humans and environments with flowing lava and volcanic ash that is a cloud of rock, mineral fragments, and glass shards.

In this part, students use latitude and longitude coordinates to map the locations of 20 active volcanoes and 20 large earthquakes around the world. Using USGS online resources, students analyze the data and observe that many of these events occur in a pattern on the edges of the Pacific Ocean.

Where do earthquakes and volcanoes occur?

PART 1

Earth materials form natural structures called landforms.

Earthquakes and volcanoes cause changes to landforms.

Most earthquakes and volcanoes occur near the edges of continents.

Events occur on land and in the ocean.

Earthquakes vary in magnitude and occur every day on Earth.

Movement of tectonic plates result in the construction of new landforms and the deconstruction of older ones.

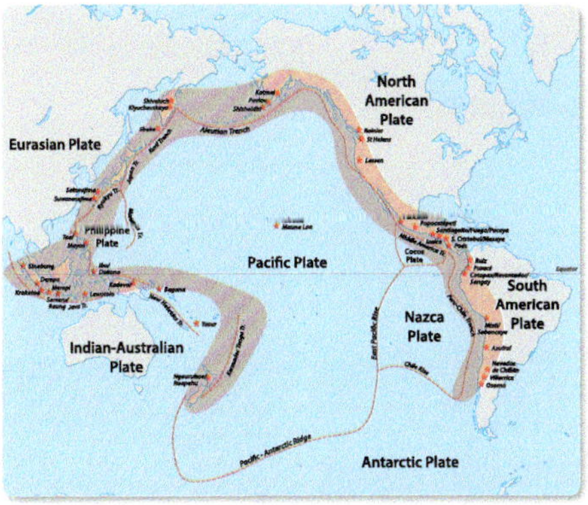

PART 1 Mapping Earthquakes and Volcanoes

Materials

PROVIDED EQUIPMENT

For each student

1 *FOSS Science Resources: Soils, Rocks, and Landforms* "Events that Change Landforms" (optional)

SCHOOL-SUPPLIED ITEMS

For each student

2 Pencils, each a different color (blue and red)

1 Ruler or other straight edge (optional)

For the class

- Poster: "This Dynamic Planet" (download front side from USGS for projection)

FOSSWEB RESOURCES

For each student

- Teacher master 11, *Volcanoes and Earthquakes Data Sheet*
- Teacher master 12, *World Map*

For the class

- Word wall cards and equipment cards
- Teaching slides
- Anchor Phenomenon Resource Video 3

For the teacher

- *Embedded Assessment Notes*

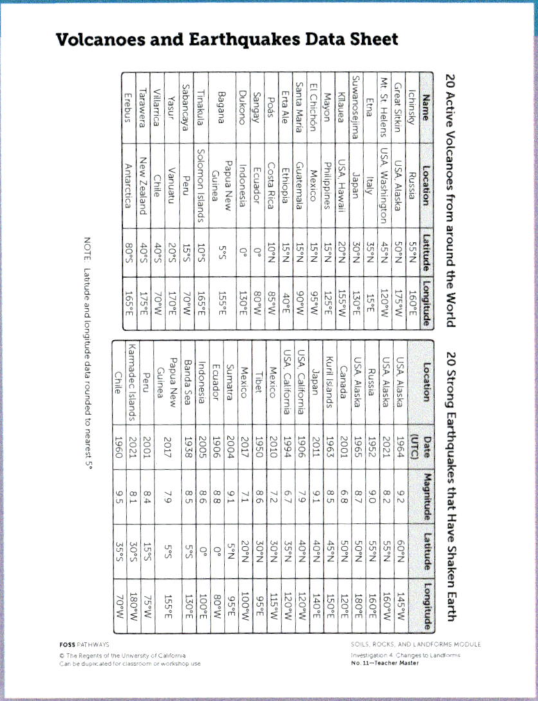

Volcanoes and Earthquakes Data Sheet

20 Active Volcanoes from around the World

Name	Location	Latitude	Longitude
Ichinsky	Russia	55°N	160°E
Great Sitkin	USA, Alaska	50°N	175°W
Mt. St. Helens	USA, Washington	45°N	120°W
Etna	Italy	35°N	15°E
Suwanosejima	Japan	30°N	130°E
Kīlauea	USA, Hawaii	20°N	155°W
Mayon	Philippines	15°N	125°E
El Chichón	Mexico	15°N	95°W
Santa María	Guatemala	15°N	90°W
Erta Ale	Ethiopia	15°N	40°E
Poás	Costa Rica	10°N	85°W
Sangay	Ecuador	0°	80°W
Dukono	Indonesia	0°	130°E
Bagana	Papua New Guinea	5°S	155°E
Tinakula	Solomon Islands	10°S	165°E
Sabancaya	Peru	15°S	70°W
Yasur	Vanuatu	20°S	170°E
Villarrica	Chile	40°S	70°W
Tarawera	New Zealand	40°S	175°E
Erebus	Antarctica	80°S	165°E

20 Strong Earthquakes that Have Shaken Earth

Location	Date (UTC)	Magnitude	Latitude	Longitude
USA, Alaska	1964	9.2	60°N	145°W
USA, Alaska	2021	8.2	55°N	160°W
Russia	1952	9.0	55°N	160°E
USA, Alaska	1965	8.7	50°N	180°E
Canada	2001	6.8	50°N	120°E
Kuril Islands	1963	8.5	45°N	150°E
Japan	2011	9.1	40°N	140°E
USA, California	1906	7.9	40°N	120°W
USA, California	1994	6.7	35°N	120°W
Mexico	2010	7.2	30°N	115°W
Tibet	1950	8.6	30°N	95°E
Mexico	2017	7.1	20°N	100°W
Sumatra	2004	9.1	5°N	95°E
Ecuador	1906	8.8	0°	80°W
Indonesia	2005	8.6	0°	100°E
Banda Sea	1938	8.5	5°S	130°E
Papua New Guinea	2017	7.9	5°S	155°E
Peru	2001	8.4	15°S	75°W
Karmadec Islands	2021	8.1	30°S	180°W
Chile	1960	9.5	35°S	70°W

NOTE: Latitude and longitude data rounded to nearest 5°

FOSS PATHWAYS

SOILS, ROCKS, AND LANDFORMS MODULE
Investigation 4: Changes to Landforms
No. 11—Teacher Master

No. 11—Teacher Master

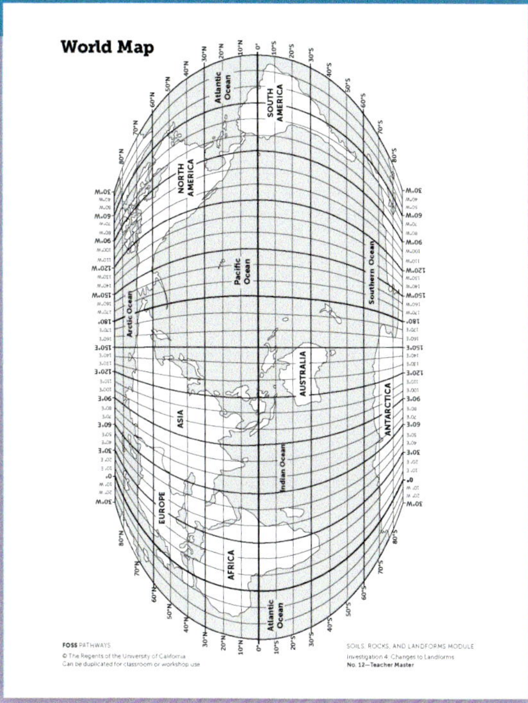

No. 12—Teacher Master

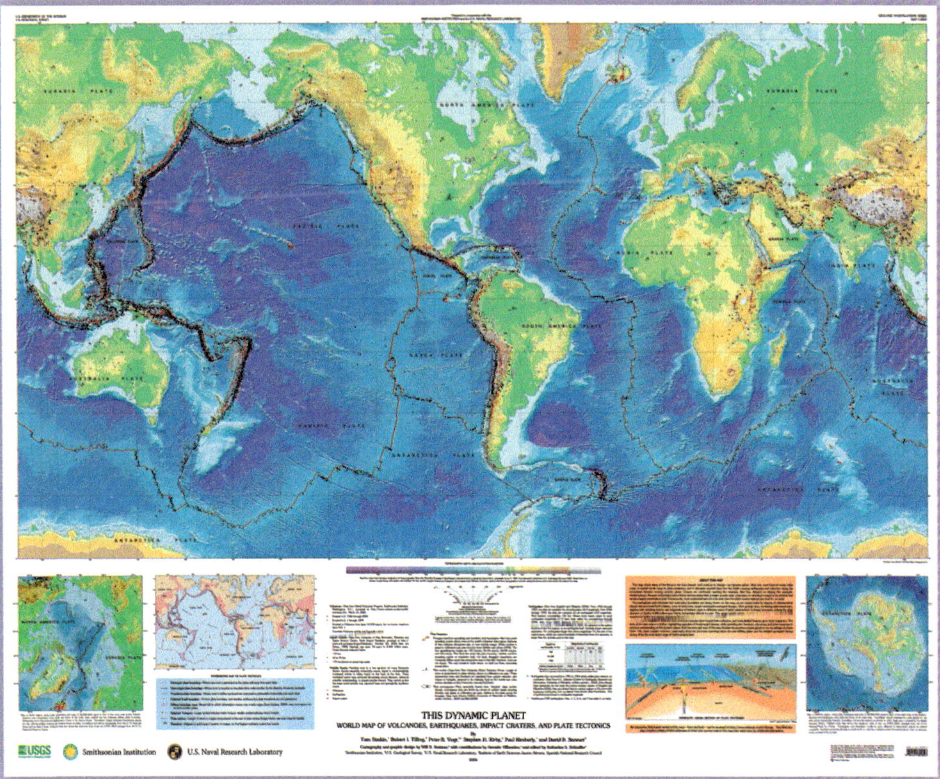

USGS This Dynamic Planet (front side)

INVESTIGATION 4

PART 1 Mapping Earthquakes and Volcanoes

Getting Ready

Schedule

2 Sessions
Session 1: Steps 1–10
Session 2: Steps 11–21 (Steps 19 and 22 optional)

Review

- Background for Part 1
- FOSSweb Resources for this part
- Optional reading: "Events that Change Landforms"
- Online activity: "Latest Earthquakes" (USGS interactive map)
- Online poster: "This Dynamic Planet" (USGS)
- Video: *Earthquakes and Volcanoes*
- Video: *Shake Alert: Earthquake Early Warning Systems for the West Coast of the United States*
- Optional video: *Remembering Ridgecrest*
- Teacher Preparation Video
- Anchor Phenomenon Resource Video 3

Print or Copy

For each student:

- Teacher master 11, *Volcanoes and Earthquakes Data Sheet*
- Teacher master 12, *World Map*

For the teacher:

- *Embedded Assessment Notes*

Prepare Material

- Plot volcanoes and earthquakes A
- Find local coordinates B
- Prepare settings for "Latest Earthquakes" USGS interactive map C

Plan for Assessment

- Review Step 18, "What to Look For" in the notebook entry

PREPARATION DETAILS

A Plot volcanoes and earthquakes

Prepare your copy of volcano and earthquake distributions on teacher master 12, *World Map*. Use the data on teacher master 11, *Volcanoes and Earthquakes Data Sheet*, to plot the occurrences of 20 volcanoes and 20 earthquakes around the world. Differentiate between volcanoes and earthquakes by plotting the volcanoes with red triangles and earthquakes with blue dots (see Step 8).

B Find local coordinates

Find the coordinates (latitude and longitude) of your city's location. A simple Internet search of your city's name and the word "coordinates" will yield the information. Simplify to whole number degrees. For example, the results for Chicago are:

Latitude: 41.8781°N; simplify to 42°N
Longitude: 87.6298°W; simplify to 88°W

C Prepare settings for "Latest Earthquakes"

Go to the website USGS Earthquake Hazards Program and click on "Latest Earthquakes" for the interactive map.

a. Zoom out until you have the entire world map on the screen.

b. Click on the "Layers" icon (upper right) to select Terrain and Plate Boundaries.

c. Click on the "Key" icon for magnitude and age representations.

d. Click on the "Gear" icon for age settings. Click through each one and then set it to what you want to start with in class. The age settings are open on the right on the display below.

e. For more details on how to use the interactive map, click on "?"

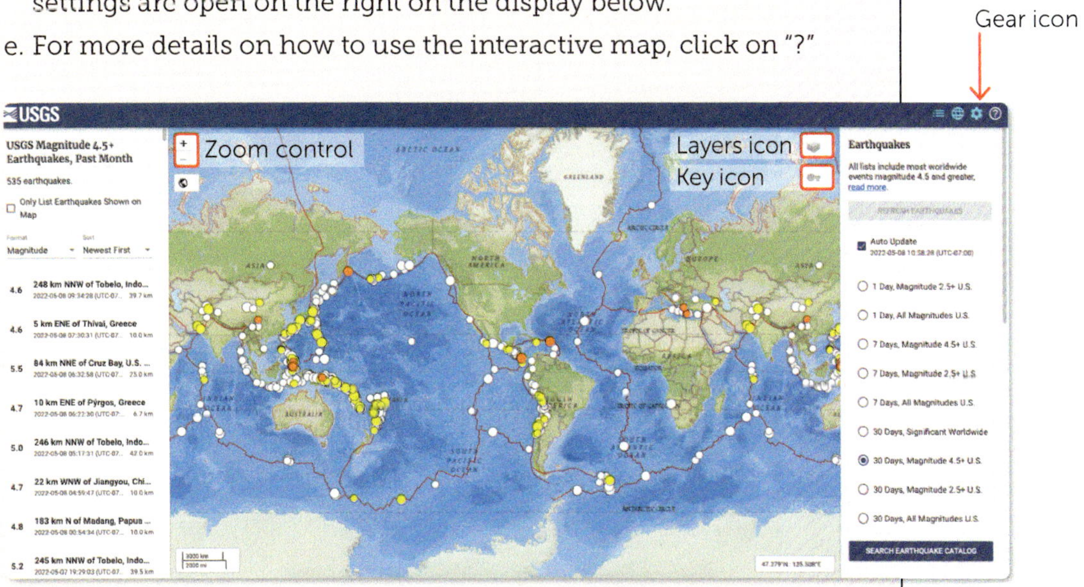

PART 1 Mapping Earthquakes and Volcanoes

Guiding the Investigation

SESSION 1 **45 MINUTES**

1 Introduce the anchor phenomenon—Earth-shaking events

Have students share what they know about volcanoes and earthquakes from previous activities in this module. Students can refer to the images in the *FOSS Science Resources* book, "Events that Change Landforms"—the images of Mount St. Helens' eruption and damage to structures from earthquakes. Give students a chance to describe any personal experience with Earth-shaking events.

Science and Engineering Practices

- Asking questions

Show the Anchor Phenomenon Resource Video 3 or describe the scenario:

Students and their teacher in a classroom in Alaska dive under their desks when the building starts shaking. They wait until the vibrations stop and quickly evacuate the school.

Ask students to share what they observed in the video and ask questions about the event.

2 Introduce the focus/driving question

NOTE

The driving question and the focus question are the same in this part.

FOCUS QUESTION

Where do earthquakes and volcanoes occur around the world?

Project or write the focus question on the board while you read it aloud.

➜ *Where do earthquakes and volcanoes occur around the world?*

Have students write the date and question in their notebooks. Tell them that they will be mapping data where some of these events occurred on a world map.

3 Introduce the world map

Distribute a copy of teacher master 12, *World Map*, to each student. Project a copy of this teacher master and ask,

➜ *What do you notice about the map?* [There are both horizontal and vertical lines on the map with numbers at the end of the lines. In the center of the map is the Pacific Ocean. It is an oval projection rather than rectangular.]

4 Introduce *latitude*

Point out the equator on the projected map. Have students locate the equator on their map. Ask them to trace the bold equator line left or

right to notice its 0 degrees notation. Point out the 10 degree intervals north and south of the equator. Ask,

→ *How might these numbers help you to use this map?* [It provides a grid to locate a point north or south of the equator.]

Provide the north/south **latitude** number for your city's location. Ask,

→ *What do you think the lines above and below the equator are called?* [The lines north and south of the equator are called lines of latitude. One way to remember that term is to think about a ladder with horizontal rungs.]

Together find the latitude line for your location on the projected map and have the students locate it on their map.

5 Introduce *longitude*

Ask,

→ *Now that we know our latitude, do we have enough information to locate our city on the map?* [No, we need to know how far east or west our location is on that latitude line.]

Explain that the lines that run vertically are called lines of **longitude**.

Have students locate the line marked 180° roughly in the middle of the map. Identify this as the International Date Line. Then have the students locate the line marked 0° far to the left. Identify this as the prime meridian. Explain that this line goes through Greenwich, England, and is the arbitrary, but accepted, imaginary line that divides the world into east and west from the North Pole to the South Pole.

6 Get oriented with coordinates

Point out to students that because this map is centered on the Pacific Ocean, the eastern lines of longitude are on the left-hand side of the map and the western ones are on the right-hand side. This is the reverse of the conventional west/east orientation.

Provide the east/west longitude number for your location. Together locate your city with both **coordinates**. Model using one finger to mark the latitude and using another finger to mark longitude. Then carefully pull your fingers together to locate where the two numbers intersect. Put a star in the approximate location of your city.

Another method is to use a piece of paper or ruler to mark the line of latitude and then trace the line of longitude down to the ruler and find the intersection of the numbers that way.

Explain that the latitude and longitude together provide the set of coordinates in **degrees**. Every location in the world can be located with a set of coordinates.

ENGLISH LANGUAGE SUPPORT
Make a chart with a simple map and add the words in bold and others that may be new to students to use as a reference.

PART 1 Mapping Earthquakes and Volcanoes

SESSION 1

7 Introduce the data sheet

Tell students that now they have a system to map the locations of volcanoes and earthquakes. Now they need data. Distribute a copy of teacher master 11, *Volcanoes and Earthquakes Data Sheet*. Give students time to review the organization of the data. They should review the column headings for both the volcano and earthquake data. Explain,

An earthquake location is given based on its ***epicenter****. The epicenter is the surface point on Earth above where the underground rock in the crust breaks.*

Ask them if they have any questions about the data.

Science and Engineering Practices

- Asking questions

NOTE

You might want to have one pair of students in each group work together to plot the earthquakes while the other pair of students in each group work together to plot the volcanoes. Then they can switch so all students plot both kinds of events.

8 Establish a legend

Challenge students to use the coordinate system to map the location of the 20 volcanoes and 20 earthquakes. Suggest establishing colors and symbols to visually display the two kinds of events. Students should use one color of pencil for earthquakes (blue) and another color for volcanoes (red). In addition, they should have different shapes to mark the location of the two kinds of event. Have them use a triangle for each volcano (red triangle) and a circle for each earthquake (blue circle).

Science and Engineering Practices

- Developing and using models

9 Plot a few events as a class

Plot the first earthquake and volcano data points together to make sure that students are confident in their understanding of the modeling system. Here's a process you might use.

- The first earthquake listed happened in Alaska in the United States in 1964. The latitude and longitude coordinates are 60°N latitude and 145°W longitude.
- Start by finding the latitude (ladder) line and place your finger on that line. Be sure that you have your finger on the 60°N line.
- Now look for the longitude line and place a finger from your other hand on that line. Again, be sure you have your finger on 145°W.
- Drag your two fingers together along their lines until they meet. Use your blue pencil to mark a circle at the point where your fingers meet. This is the location, or epicenter, of that earthquake. You have plotted your first point.
- Let's do another example using a volcano location. The first volcano listed is the Ichinsky volcano in Russia. Its coordinates are 55°N latitude and 160°E longitude.

- Plot this volcano just like we did the earthquake. Find the 55°N latitude line and place your finger on it. Then find the 160°E longitude line and place a different finger on that line. Drag your fingers toward each other on the lines until they meet. Use your red pencil to mark a triangle at the spot where they meet.

10 Map the volcanoes and earthquakes

Have students plot the remaining volcanoes and earthquakes. When they have plotted all the data, they will have 20 blue circles (earthquakes) and 20 red triangles (volcanoes) plotted on each map.

SESSION 2 **45 MINUTES**

11 Analyze and interpret the data

After students have finished mapping the events, have them analyze the data in their groups and make a list of observations and questions. Then have a short class discussion to end this session.

→ *Do you notice any patterns in the distribution of earthquakes and volcanoes around the world?* [There seems to be a lot of earthquakes and volcanoes around the Pacific Ocean.]

→ *Why do you think earthquakes and volcanoes occur in this pattern?* [Students have read in *FOSS Science Resources* about the formation of the Cascade Mountains with forces caused by sliding **tectonic plates**. Review the description of crustal plate movement called **subduction** and the questions on page 58.]

Science and Engineering Practices

- Analyzing and interpreting data
- Constructing explanations

Crosscutting Concepts

- Patterns

12 Project the USGS "Latest Earthquakes" interactive map

Introduce the USGS Earthquake Hazards Program website and show students how to navigate to the "Latest Earthquakes" interactive map. Orient students to the USGS world map and have them compare it to the world map they used to plot earthquakes and volcanoes. Ask,

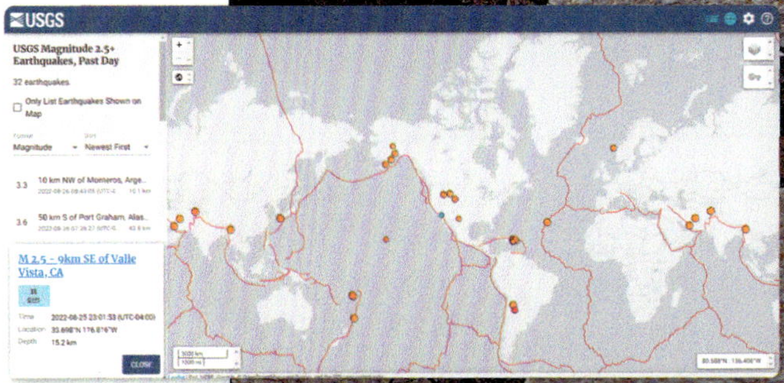

→ *What questions do you have about this map?* [They may ask what the red outlines are. You can identify those as plate boundaries and use the layers to toggle them on and off.]

Display the setting of 30 days and all magnitudes and explain what that setting means. Ask students to analyze the data and share their observations. Have them ask questions that can be answered by changing the settings of age and magnitude. It is possible to call up the data for the past 1 day, 7 days, or 30 days and to filter for the magnitude of the earthquake.

PART 1 Mapping Earthquakes and Volcanoes

Science and Engineering Practices

- Developing and using models
- Obtaining, evaluating, and communicating information

Crosscutting Concepts

- Patterns

coordinates
degrees
epicenter
latitude
longitude
subduction
tectonic plate

SESSION 2

➜ *How does the pattern of earthquakes that you plotted compare to what you see on the USGS earthquake map?* [On both maps, there is a lot of earthquake activity along the edges of the Pacific Ocean where it comes in contact with land.]

➜ *What does the USGS map show that your map doesn't?* [The crustal and oceanic plates that move and cause earthquakes. The Pacific Ocean is largely one plate.]

Offer students the term "Ring of Fire" to describe this roughly horseshoe shape where 75 percent of the active volcanoes are located. It is also where about 90 percent of the world's earthquakes occur.

13 Project the USGS "This Dynamic Planet" poster

Introduce this poster—a world map that shows both volcanoes and earthquakes. Volcanoes are red triangles and earthquakes are black dots and circles. This map includes recorded data back to 1900. It isn't possible to interpret all the data but it is possible to observe the general pattern over more than 100 years. Ask students to compare this map to their map. Have them offer their questions.

14 View the video

Have students view the video, *Earthquakes and Volcanoes*. Have them listen for new information about where and when these events occur around the world and the reasons for the pattens observed.

Give students a moment to share their observations and questions with a partner. Ask students to imagine they live near a volcano. What would make them feel safe? This is a good time to look back at the "Landforms on Display" article in the *FOSS Science Resources* book. Have students reread and discuss the photographs on pages 32–35. Use the discussion questions and prompts to lead a whole group discussion.

15 Review vocabulary

Review the new vocabulary introduced in this part including vocabulary from the video. If these words are not on the word wall, add them now.

16 Have a sense-making discussion and connect to the anchor phenomenon

Focus on volcano activity. Ask,

➜ *What is a volcano?* [The landform created by melted rock pouring or exploding from Earth's crust as lava.]

- *Why do you think some areas have volcanoes and others do not?*
- *What were some of the clues that scientists looked for to help them predict an eruption for Mount St. Helens?* [Students may mention several things they remember from the video, including the bulging of the mountaintop and increased earthquake activity.]

Next, focus on earthquake activity. Ask,

- *What is an earthquake?* [An earthquake is when the ground shakes or moves suddenly.]
- *What new, if any, patterns do you notice as you observe the volcano and earthquake locations?* [Most volcanoes and earthquakes happen along the edges of tectonic plates or along the edges of continents.]
- *Do earthquakes happen only where there are volcanoes?* [Earthquakes of different magnitudes happen every day and in places where there are no volcanoes. But the earthquakes are often associated with volcano activity. Earthquakes are more common than volcano eruptions.]

Students should notice that some earthquakes occur in areas where there are no volcanoes. Ask students to think about some reasons why volcanoes and earthquakes might occur at the same place and why not all earthquakes seem to happen where there are volcanoes.

17 Respond to the focus/driving question

Have students respond to the focus/driving question in their notebooks. They should use the information they discovered from plotting earthquakes and volcanoes as part of their responses.

- *Where do earthquakes and volcanoes occur around the world?*

18 Assess progress: notebook entry

Have students turn in their notebooks open to the page where they've constructed explanations to the focus question. Record your notes on a copy of *Embedded Assessment Notes.*

WHAT TO LOOK FOR

- *By using maps to analyze the locations of earthquakes and volcanoes, we found that these events most often occur where two or more plates meet. Plate boundaries can be found along the edges of continents.*
- *There is a pattern of earthquake and volcanic activity that surrounds the Pacific Ocean, called the Ring of Fire. Most of the earthquakes and volcanoes are in this area.*
- *Earthquakes can occur when a volcano erupts.*

ENGLISH LANGUAGE SUPPORT

For students who need scaffolding, provide a sentence frame such as:

When we analyzed the data by _____, we noticed _____. We found out that _____.

Science and Engineering Practices

- Analyzing and interpreting data
- Constructing explanations

Disciplinary Core Ideas

- ESS2.B: Plate tectonics and large-scale interactions
- ESS3.B: Natural hazards

Crosscutting Concepts

- Patterns

Interpreting Information

ELA CONNECTION

These suggested strategies address these English Language Arts objectives.

Interpret information presented visually, and explain how the information contributes to an understanding of the text.

Gather relevant information from experiences and print, and categorize the information.

Report on a text in an organized manner, using appropriate facts and relevant, descriptive details to support main ideas or themes; speak clearly at an understandable pace.

Science and Engineering Practices

- Obtaining, evaluating, and communicating information

SESSION 2

19 Read "Events that Change Landforms" (optional)

Students have studied some of the images in this reading in previous investigations to gather information about the impact of volcanoes and earthquakes. This is a good time to read the entire article to obtain, evaluate, and communicate information about the effects of natural Earth processes.

Before reading, have students review what they know so far about rapid changes to Earth's surface. Refer back to the Mount St. Helen's article and how the volcanic eruption changed the land around it and how it impacted humans and other life forms. Tell students that this article will describe other ways the land can be transformed rapidly—floods, earthquakes, and landslides.

Read the article aloud or have students read independently or in guided reading groups. Tell students to be looking for the ways the rapid changes are both constructive and destructive processes.

Another option is to jigsaw the article. Assign each group a section to read and discuss. Then, have each group develop a presentation as if they were a news station reporting on the event.

Tell students that catastrophic events are usually remembered for the destruction they cause in a short period of time. Often the destruction is measured in human losses—loss of life, buildings, and infrastructure. There are significant changes to the land as well. Some of the changes involve construction of new landforms. Thus catastrophic events often involve a combination of destructive and constructive outcomes.

EVENTS THAT **Change Landforms**

Some landforms are so old, you might think they've always been that way. The Sierra Nevada has been uplifting for more than 2 million years and is still rising. The Colorado River continues to carve the Grand Canyon, just as it has for more than 5 million years. Rock in the Appalachian Mountains began folding more than 480 million years ago. Most changes to Earth's surface are so slow, we can't see them happen.

But sometimes changes happen rapidly. Rapid changes affect people and landforms. Here are some examples of fast changes to Earth's surface.

Mount St. Helens erupting

65

Reflecting on a Phenomenon

20 Finalize the anchor phenomenon response—Earth-shaking events

Have students discuss how gathering and analyzing data helped them figure out where earthquakes and volcanoes occur around the world. Next, have them turn to a partner and share what they learned about the "Ring of Fire" phenomenon by looking at patterns in the data.

Ask them what questions they have about Earth-shaking events.

21 View the Shake-Alert video

Have students view the video, *Shake Alert: Earthquake Early Warning Systems for the West Coast of the United States* (4 min 30 seconds). Tell them this is a resource they can share with their family on FOSSweb.

After the video, ask students to discuss in their groups why it is important for humans to understand where volcano eruptions and earthquakes are likely to occur. After a few moments, have groups share ideas with the class.

Groups may share these ideas. You can also add to their ideas.

- Human safety depends on volcano and earthquake awareness and preparedness.
- Scientists constantly monitor Earth movements near volcanoes. Computers help with this monitoring. Earthquakes are one indication that a volcano may be building up to an eruption. Scientists can warn people in an area to evacuate.
- Earthquakes are harder to forecast because they don't have advance warning that we can detect. But there is one important fact about earthquakes that can help save lives. Earthquakes travel as waves outward from the epicenter. These waves travel through earth materials more slowly than computers can communicate. This means that earthquake-monitoring networks can prepare residents in other locations before the shocks arrive.

22 View video interviewing USGS scientist (optional)

Have students view the video, *Remembering Ridgecrest* (4 min duration). USGS scientists comment on the July 2019 earthquakes of 6.4 and 7.1 magnitude in a rural southern California region. The video captures the efforts of a diverse group of USGS scientists and how they worked with other agencies to provide public safety information. (This video is listed as a science extension activity in the Differentiation Opportunities at the end of the investigation.)

Science and Engineering Practices

- Analyzing and interpreting data

Crosscutting Concepts

- Patterns

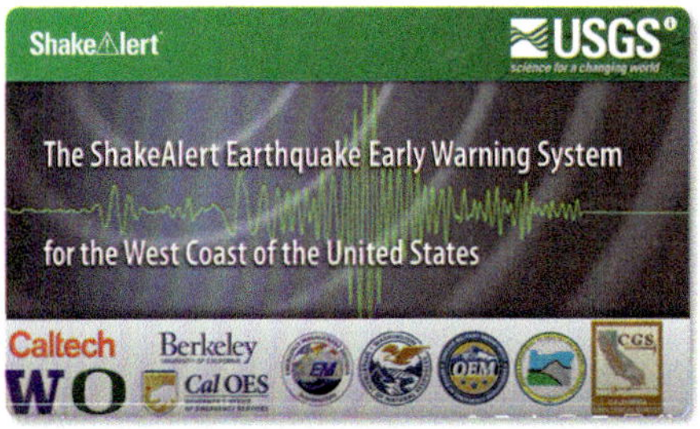

PART 2 Interpreting Rock Layers

Teacher Background

Science and Engineering Practices

- Asking questions
- Developing and using models
- Constructing explanations
- Obtaining, evaluating, and communicating information

Disciplinary Core Ideas

- ESS1.C: History of planet Earth

Crosscutting Concepts

- Patterns
- Stability and change

Phenomenon Storyline

LEARNING OBJECTIVE:

- **Students obtain information from media to construct explanations about the geological history revealed by patterns of rock layers and the presence of types of fossils.**

A family visits a natural outdoor site and finds large rock-layer formations with unusual stripes of red, pink, and yellow. Some of the layers are flat and others are tilted. In another site they discover fossils.

What do rock layers and fossils tell us about the history of a place?

Here are some things students will find out to connect to the storyline.

- Sedimentary rocks form in horizontal layers, and those layers provide information about the geological history of an area. Each layer must be older than any layer on top of it. Earthquakes can tilt those layers.
- Landforms are constantly changing due to weathering, erosion, deposition, canyon forming, and earthquakes (uplifting).
- Fossils provide evidence of organisms that lived long ago as well as clues to changes in the landscape and past environments.

FOCUS QUESTION

What Do Rock Layers and Fossils Tell Us about the History of a Place?

Most rocks on Earth's surface are sedimentary. They formed from sediment, bits of older rock that were broken apart by water or wind. Some of these rocks contain fossils.

Types of rock can tell us about the environment in which they formed.

- Sandstone forms from sediment deposited in sandy areas such as beaches, deserts, and dunes.
- Limestone forms from sediment deposited in water, mostly in marine environments like the ocean.
- Shale forms from sediment deposited in calm, muddy environments like swamps.

Sedimentary rocks form particle by particle and layer by layer. Gravel, sand, and mud settle on the bottoms of rivers, lakes, and the ocean. The layers pile up, one on top of the other. This process (called superposition) provides an important clue to the relative ages of rock layers and the fossils in them. Each layer must be older than any layer on top of it.

Aa

shale

Horizontal layers form when sediment settles from water or air. But many layered rocks are no longer horizontal. Why is that? The rocks moved from their original horizontal position. Events such as mountain formation and earthquakes can fold, bend, and tilt sedimentary rock layers.

Rock layers are called strata. Stratigraphy is the study of the distribution and sequence of sedimentary rock layers. If two rock layers, or strata, contain the same group of fossils, they must be the same age. These ideas are important in using fossils to determine the relative age of rocks.

- Fossils represent the remains, traces, or imprints of organisms that once lived.
- Most fossils represent the remains of organisms that are now extinct. They come from species that are no longer living.
- The structure of fossils indicates the ancient environment at the time of deposition.

The most powerful way to analyze sedimentary rocks is to look at the order and thickness of the layers in relation to each other. Layers of sedimentary rocks can show the changes in the environment of a location. They record in rock the rising and lowering of the ocean, changes in climate, changes in elevation, periods of deposition, periods of erosion, and periods of uplifting or downcutting.

In the last part, students gather information through case studies of sedimentary rock formations. In one case study, slanted rock in sandstone layers expose the effect of tremendous earthquakes or tectonic forces that can bend, twist, break, and uplift rock layers to expose them in an otherwise flat area. A second case study reveals trilobite fossils in sedimentary shale providing evidence that this area was once a marine coastal environment. The mystery is how the shale ended up forming a mountain in the middle of a desert. Several rock-layer puzzles provide opportunities for students to construct explanations integrating many concepts developed throughout the module.

Where do earthquakes and volcanoes occur?

PART 1

Earth materials form natural structures called landforms.

Earthquakes and volcanoes cause changes to landforms.

Most earthquakes and volcanoes occur near the edges of continents.

- Events occur on land and in the ocean.
- Earthquakes vary in magnitude and occur every day on Earth.

Movement of tectonic plates result in the construction of new landforms and the deconstruction of older ones.

PART 2

What do rock layers tell us about the history of a place?

Some events happen over long periods of time.

Landforms and environments change. The sequence of horizontal rock layers record changes.

- Fossils in layers indicate past environments.

Weathering, erosion, deposition, uplifting, and downcutting change landforms.

INVESTIGATION 4

PART 2 Interpreting Rock Layers

Materials

PROVIDED EQUIPMENT

For each student

1 *FOSS Science Resources: Soils, Rocks, and Landforms*
"Valley of Fire"
"Fossils Tell a Story"
"Frenchman Mountain"

For the class

10 Banded sandstone rock samples
10 Limestone rock samples
10 Shale rock samples
10 Granite rock samples
8 FOSS trays

FOSSWEB RESOURCES

For each student

- Notebook sheet 18, *Rock-Layer Age Puzzle*
- *Posttest*

For the class

- Word wall cards and equipment cards
- Teaching slides
- Anchor Phenomenon Resource Video 4

For the teacher

- *Embedded Assessment Notes*
- *Assessment Record*

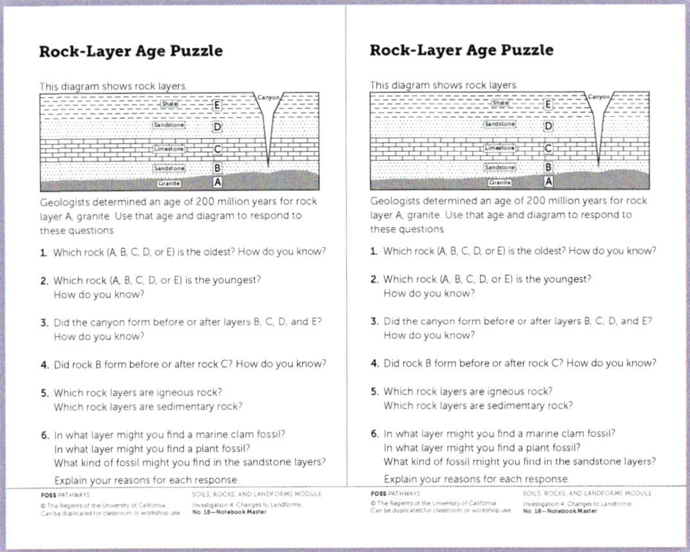

Rock-Layer Age Puzzle

This diagram shows rock layers.

Geologists determined an age of 200 million years for rock layer A, granite. Use that age and diagram to respond to these questions.

1. Which rock (A, B, C, D, or E) is the oldest? How do you know?
2. Which rock (A, B, C, D, or E) is the youngest? How do you know?
3. Did the canyon form before or after layers B, C, D, and E? How do you know?
4. Did rock B form before or after rock C? How do you know?
5. Which rock layers are igneous rock? Which rock layers are sedimentary rock?
6. In what layer might you find a marine clam fossil? In what layer might you find a plant fossil? What kind of fossil might you find in the sandstone layers? Explain your reasons for each response.

FOSS PATHWAYS
© The Regents of the University of California. Can be duplicated for classroom or workshop use.
SOILS, ROCKS, AND LANDFORMS MODULE
Investigation 4: Changes to Landforms
No. 18—Notebook Master

Rock-Layer Age Puzzle

This diagram shows rock layers.

Geologists determined an age of 200 million years for rock layer A, granite. Use that age and diagram to respond to these questions.

1. Which rock (A, B, C, D, or E) is the oldest? How do you know?
2. Which rock (A, B, C, D, or E) is the youngest? How do you know?
3. Did the canyon form before or after layers B, C, D, and E? How do you know?
4. Did rock B form before or after rock C? How do you know?
5. Which rock layers are igneous rock? Which rock layers are sedimentary rock?
6. In what layer might you find a marine clam fossil? In what layer might you find a plant fossil? What kind of fossil might you find in the sandstone layers? Explain your reasons for each response.

FOSS PATHWAYS
© The Regents of the University of California. Can be duplicated for classroom or workshop use.
SOILS, ROCKS, AND LANDFORMS MODULE
Investigation 4: Changes to Landforms
No. 18—Notebook Master

No. 18—Notebook Master

PART 2 Interpreting Rock Layers

Getting Ready

Schedule

5 Sessions

Session 1: Steps 1–5
Session 2: Steps 6–7
Session 3: Step 8–11
Session 4: Step 12–18
Session 5: Step 19, *Posttest*

Review

- Background for Part 2
- FOSSweb Resources for this part
- Readings: "Valley of Fire," "Fossils Tell a Story," and "Frenchman Mountain"
- Online activity: "Rock-Layer Movie Maker"
- Teacher Preparation Video
- Anchor Phenomenon Resource Video 4

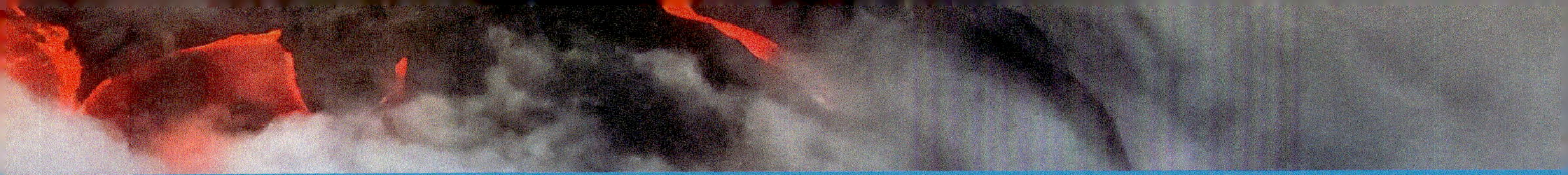

Print or Copy

For each student:

- Notebook sheet 18, *Rock-Layer Age Puzzle*

For the teacher:

- *Embedded Assessment Notes*
- *Assessment Record*

Prepare Material

- Prepare rock samples in trays ______ A
- Plan for devices ______ B

Plan for Assessment

- Review Step 16 "What to Look For" in the notebook entry
- Review Step 19 in Guiding to prepare for the *Posttest*

PREPARATION DETAILS

A Prepare rock samples in trays

Prepare a tray of labeled rock samples for each group. Start by making a set of four labels (granite, banded sandstone, limestone, shale) for each tray using self-stick notes. One rock label goes in each large depression of a tray. Place the corresponding rock sample on top of each label.

B Plan for devices

Plan to have enough devices for students to work individually or in pairs. Test the functionality of "Rock-Layer Movie Maker" on the students' devices. The activity builds on students' understanding of ages of horizontal rock layers or sedimentary rocks and the formation of fossils in those layers developed through the readings.

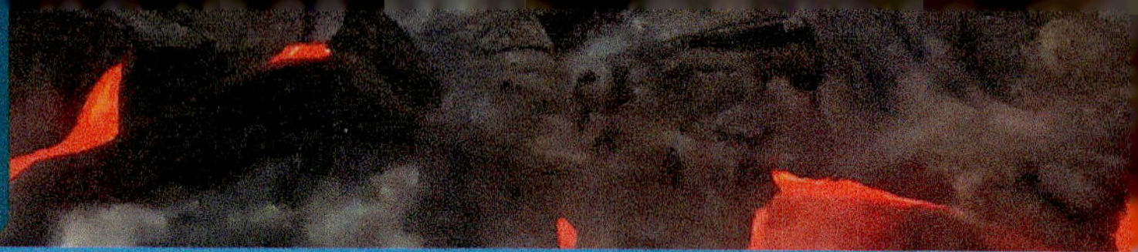

PART 2 Interpreting Rock Layers

Guiding the Investigation

NOTE
This is a sample anchor phenomenon and can be customized to provide local relevance.

SESSION 1 **45 MINUTES**

1 Introduce the rock-layers flat and tilted phenomenon

Begin this part by sharing the scenario and then showing the Anchor Phenomenon Resource Video 4.

A family visits a natural outdoor site and finds large rock-layer formations with unusual stripes of red, pink, and yellow. Some of the layers are flat and others are tilted. In another site they discover fossils.

After viewing the Anchor Phenomenon Resource Video 4, ask students to share their observations and questions.

Ask students what information they need to help them figure out what happened in this place. Have students share their thinking with a partner.

Science and Engineering Practices
- Asking questions

Interpreting Information

2 Read "Valley of Fire"

Have students read the article to obtain, evaluate, and communicate information about sedimentary rock formation. Remind students of the rock-layer phenomenon they observed in the video and the scenario of the family who observed these large rock formations with colorful stripes and tilted layers. Tell students that they will read about how members of the family figured out how these landforms came to be. Give students a moment to preview the text by looking at and discussing the photographs, diagram, and captions with a partner. To heighten engagement, you might assign different students to read aloud the parts of the narrator, Savannah, her dad, and her mom. This can be done whole-group or in groups of four. Pause to discuss the questions and prompts. Encourage students to site examples and details from the text to support their ideas.

ELA CONNECTION

These suggested strategies address these English Language Arts objectives.

Refer to details and examples in a text when explaining what the text says explicitly and when drawing inferences from the text.

Read and comprehend informational science texts.

Science and Engineering Practices

- Obtaining, evaluating, and communicating information

Activate Prior Knowledge Ask students to share any experiences they have had exploring these types of rock formations and/or what questions they have about them.

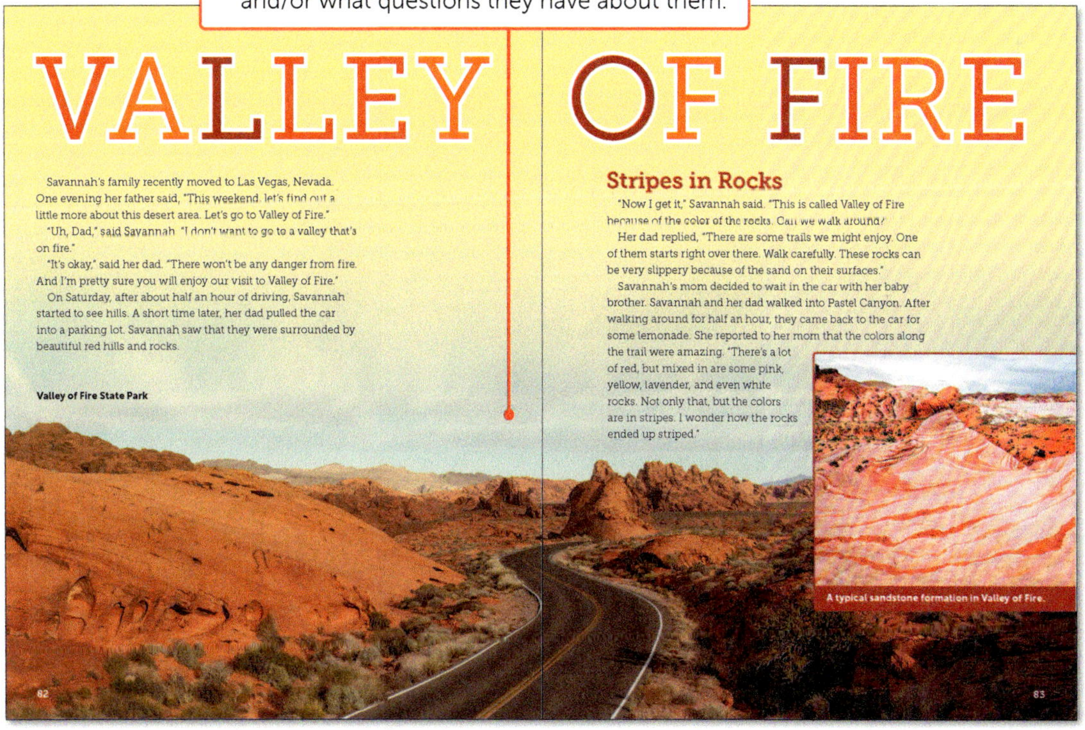

VALLEY OF FIRE

Savannah's family recently moved to Las Vegas, Nevada. One evening her father said, "This weekend, let's find out a little more about this desert area. Let's go to Valley of Fire."

"Uh, Dad," said Savannah. "I don't want to go to a valley that's on fire."

"It's okay," said her dad. "There won't be any danger from fire. And I'm pretty sure you will enjoy our visit to Valley of Fire."

On Saturday, after about half an hour of driving, Savannah started to see hills. A short time later, her dad pulled the car into a parking lot. Savannah saw that they were surrounded by beautiful red hills and rocks.

Valley of Fire State Park

Stripes in Rocks

"Now I get it," Savannah said. "This is called Valley of Fire because of the color of the rocks. Can we walk around?"

Her dad replied, "There are some trails we might enjoy. One of them starts right over there. Walk carefully. These rocks can be very slippery because of the sand on their surfaces."

Savannah's mom decided to wait in the car with her baby brother. Savannah and her dad walked into Pastel Canyon. After walking around for half an hour, they came back to the car for some lemonade. She reported to her mom that the colors along the trail were amazing. "There's a lot of red, but mixed in are some pink, yellow, lavender, and even white rocks. Not only that, but the colors are in stripes. I wonder how the rocks ended up striped."

A typical sandstone formation in Valley of Fire.

82 83

PART 2 Interpreting Rock Layers

SESSION 1

Develop Understanding Have students take turns paraphrasing how sedimentary rock formed with a partner.

How the Striped Rocks Formed

Her dad suggested they look at the state park exhibit at the start of the trail. After studying the exhibit for a while, they returned to the car. Her mom asked Savannah if the information helped her understand the colorful stripes.

Savannah responded, "It's starting to make sense. These rocks are all sandstone. Sandstone is a sedimentary rock that forms when sand gets buried under other sediment for a very long time. Minerals soak into the sand. Weight and pressure increase as more and more layers of sediment pile up on top of each other. After maybe 100 million years, the sand changes into sandstone."

Savannah continued, "All the sand that turned into sandstone was deposited a little bit at a time in layer after layer. Each new layer of sand was a little different than the sand that was deposited before it. Each new layer of sand might be the same color as the sand below it or a different color. Even when the sandstone was the same color, if you look closely, you can see that the sand was deposited in separate layers."

Sand Dunes and Wind

Savannah's mom asked, "What are the layers called?"

Savannah replied, "The exhibit called the layers **strata**. The process of placing sand particles into a stratum is deposition."

Savannah's dad said, "Sand layers are deposited by moving water or wind. At Valley of Fire, the sandstone formed from ancient sand dunes that covered the area during the Jurassic Period from about 150–200 million years ago. Over a very long time, wind deposited many layers. The layers started out flat and horizontal. The first horizontal layer of sand was like a parking lot, and the next layer of sand was deposited on top of the first layer like new pavement on a parking lot. This continued for millions of years. Because the sand started out as layers, what we see as stripes is actually the edges of the layers of colored sand that turned into sandstone."

How did these different rock layers form in the sandstone?

84 85

Rock Layers at Angles

Then her dad said, "There's one more rock formation I think we should take a look at before we head for home." The family drove a short distance and turned into another parking lot. They walked to an overlook point.

Her dad asked, "What do you see here?"

Savannah saw what looked like a sandstone hill with colorful strata at strange angles. She said, "That hill has strata that were deposited at slanted angles. How did that happen? I thought that sentiments settled in horizontal layers on top of each other. Could it be that those layers of rock started out flat and horizontal, and then something happened to tip them sideways?"

"Well, now you have a mystery to solve," said her dad. "You're probably not the first person to ask that question. You can do some research when we get home."

Colorful, slanted rock strata at Valley of Fire

Can you see the different colors and patterns of strata on the rocks?

Uplifting Forces

At home, Savannah used the family computer to search for the cause of rock tilting. The thing that interested her most was about the **tectonic forces** at work inside the Earth. These forces can lift and break layers of rock near the surface. The forces can change the orientation of flat layers of rock. The rock layers became tilted or twisted.

One article explained it this way. Place two books flat on a table with their open edges touching. The covers and pages represent a formation of rocks made of many layers of deposited sediment. All the layers (pages) are horizontal and flat. Then, imagine a force pushing up between the two books. The force breaks the formation down the middle, pushes the two halves apart, and tilts them up on their bindings. The pages (layers) are still flat, but they are no longer horizontal. The rock layers that were originally horizontal are now slanted.

After Savannah explained this to her dad, he said, "That's interesting. We never think of rock as a material that can be bent, twisted, and broken by strong forces. But it does explain the tilted strata we saw in the rocks in Valley of Fire."

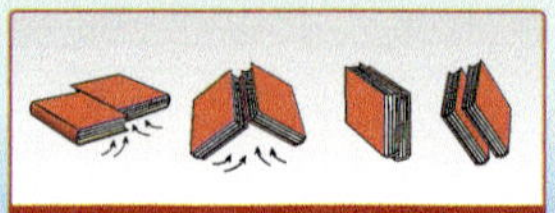

Using books to model how tectonic forces can break and tilt rock layers

Questions for Reflection

1. What was the first thing that Savannah observed about the rocks and what question did she ask?
2. What did she learn that helped her to answer her question?
3. Which rock layer is older, the one at the top or the one at the bottom? How do you know?
4. How do you explain rock layers that are tilted?

86 87

Apply Understanding Have students act out the process of tectonic forces using two books as shown in the diagram or with their hands. One student can read the explanation as the other acts it out.

3 Discuss the reading and sense-making

Guide a whole-group discussion using the Questions for Reflection.

- *What was the first thing that Savannah observed about the rocks and what question did she ask?* [She noticed the bright colors and the stripes. She asked, how did the rocks end up striped?]
- *What did she learn that helped her to answer her question?* [That the rocks were sandstone, a sedimentary rock that forms when sand is buried under other sediment for millions of years. Each layer of sand is deposited on top of the previous layer, and in some cases, the sand that forms a layer might be a different color from the previous sand layer. Layers form by moving water or wind.]
- *Which rock layer is older, the one at the top or the one at the bottom? How do you know?* [The layer of sandstone at the bottom is the oldest because it was deposited first and the layer at the top is the youngest as it was deposited and turned to rock last.]
- *How do you explain rock layers that are tilted?* [Rock layers are formed in flat, horizontal layers so something happened after the layers were formed to cause the layer to tilt. Strong earthquakes or tectonic forces can cause this effect—uplifting.]

4 Review how sedimentary rocks form

Suggest that students review the text and photographs in the section "Sedimentary Rocks" from the article "Where Do Rocks Come From?" in their *FOSS Science Resources* book. Have students take turns paraphrasing with a partner how sedimentary rock was formed. They should include that they form from bits and pieces of recycled rocks and minerals. These sediments form layers and are pressed and cemented together over millions of years. Sandstone formations can be very large layers; sandstone rocks break off from the layers through weathering.

5 Introduce the focus/driving question

FOCUS QUESTION

What do rock layers and fossils tell us about the history of a place?

Project or write the focus question on the board as you read it aloud. Have students write the date and focus question in their notebooks and start a response to it with their current ideas. They will continue to gather information and complete their responses in the next sessions.

- *What do rock layers and fossils tell us about the history of a place?*

ELA CONNECTION

These suggested strategies address these English Language Arts objectives.

Refer to details and examples in a text when explaining what the text says explicitly and when drawing inferences from the text.

Explain events, procedures, ideas, or concepts in a scientific text including what happened and why, based on specific information in the text.

Interpret information presented visually.

ENGLISH LANGUAGE SUPPORT

Have students begin with an explanation of rock layers. If needed, provide a writing frame such as,

Rock layers are formed when _____ . *First,* _____ . *Then,* _____ . *Next,* _____ . *I think rock layers can tell us* _____ .

NOTE

The driving question and the focus question are the same in this part.

PART 2 Interpreting Rock Layers

Science and Engineering Practices

- Obtaining, evaluating, and communicating information

ELA CONNECTION

These suggested strategies address these English Language Arts objectives.

Determine the main idea of a text and explain how it is supported by key details; summarize the text.

Interpret information presented visually, orally, or quantitatively and explain how the information contributes to an understanding of the text in which it appears.

SESSION 2 **45 MINUTES**

6 Read "Fossils Tell a Story"

Review the focus question students responded to at the end of the last session. Introduce the next reading by explaining that this article will provide more information about fossils, which are found in sedimentary rocks and rock layers.

Introduce the title and ask students to discuss what they think it means. Then, have them preview the text by looking at and discussing the photographs and reading the captions. Ask students to write down a few questions they have about fossils that might be answered in the text. Add that they should look for more information to add to their focus question response.

Read the text aloud or have students read independently. Then have students reread the text a second time. Pause after each section or paragraph and have students turn and talk to their reading partner about the main idea and any questions they have. Encourage students to visualize the processes the text is explaining and to refer to the photographs for clarification. Use the prompts and questions to guide whole-group discussions.

Develop Understanding Have students discuss how they think scientists can determine the age of Earth by digging through layers of sedimentary rock.

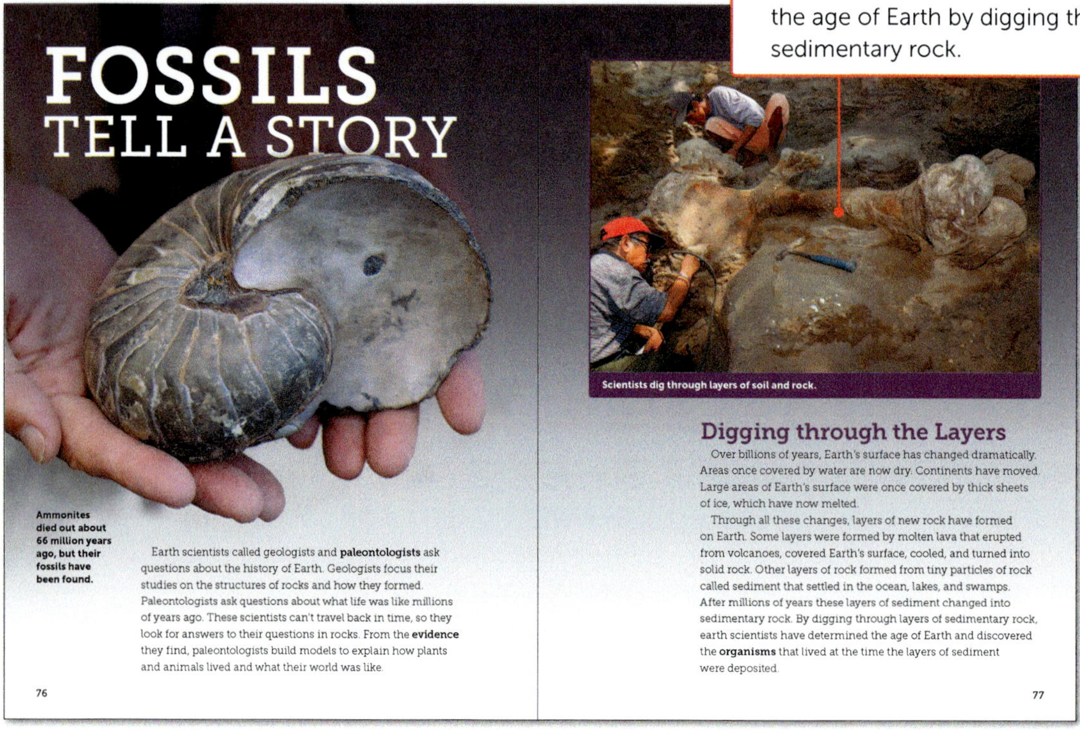

FOSSILS TELL A STORY

Ammonites died out about 66 million years ago, but their fossils have been found.

Earth scientists called geologists and **paleontologists** ask questions about the history of Earth. Geologists focus their studies on the structures of rocks and how they formed. Paleontologists ask questions about what life was like millions of years ago. These scientists can't travel back in time, so they look for answers to their questions in rocks. From the **evidence** they find, paleontologists build models to explain how plants and animals lived and what their world was like.

76

Scientists dig through layers of soil and rock.

Digging through the Layers

Over billions of years, Earth's surface has changed dramatically. Areas once covered by water are now dry. Continents have moved. Large areas of Earth's surface were once covered by thick sheets of ice, which have now melted.

Through all these changes, layers of new rock have formed on Earth. Some layers were formed by molten lava that erupted from volcanoes, covered Earth's surface, cooled, and turned into solid rock. Other layers of rock formed from tiny particles of rock called sediment that settled in the ocean, lakes, and swamps. After millions of years these layers of sediment changed into sedimentary rock. By digging through layers of sedimentary rock, earth scientists have determined the age of Earth and discovered the **organisms** that lived at the time the layers of sediment were deposited.

77

Apply Understanding Have students compare how index fossils are different from other types of fossils. Why are index fossils especially helpful to scientists?

Fossils Found

Scientists want to determine the age of the rocks they study. They use several tools. One technique is based on the fossils found in the rocks. Fossils are the remains of organisms preserved in rocks. Scientists know that some of the organisms represented by the fossils are much older than other fossils. This tells a scientist that one rock is older than another (based on the fossils found in them), but not how old either rock is in absolute terms. To determine the absolute age of rocks, geologists use other tools and techniques.

Fossils form in several different ways. Most form when a dead organism ends up at the bottom of a water environment (the ocean, a lake, or a swamp). The organism gets buried in mud and sediment. Most of the softer parts of the organism decompose, but the hard parts (bones, teeth, some shells) remain. Sediment carried by erosion continues to deposit on top of the remains. Over very long periods of time, the sediment gets compacted and eventually hardens into rock. As buried bones age, minerals seep into them and slowly turn the bones into solid rock. A bone fossil is no longer bone but a rock in the exact shape of the original bone. This is the same way a tree will turn into **petrified wood**. A piece of petrified wood is a tree fossil, preserving the tree's wood grain and growth rings exactly.

A stingray fossil

A fern fossil

78

Look at the mold and cast of an ammonite. (The mold is an impression of the ammonite while the cast looks like the actual ammonite.) Ammonites are extinct ocean animals that had spiral shells. They are excellent index fossils and are used to date the rocks.

Sometimes a buried organism decomposes. This leaves a space in the sediment where the organism was originally. The sediments around the space fossilize. This type of fossil is called a **mold**. It is an impression of the organism. Another kind of fossil is called a **cast**. To make a cast, the remains of the organism are filled with minerals, turned into rock, and form an exact replica of the original organism. Cast fossils of organisms, like clams, sea urchins, and snails, are fairly common.

After years of studying fossils found all over the planet, paleontologists have discovered that some of the fossil organisms lived on Earth for only a few million years, but are found in many places. A few are found all over the world, in the United States, Europe, Asia, and South America. Such short-lived, widespread fossils are known as **index fossils**. Index fossils help Earth scientists determine the absolute age of the rocks they are studying. When scientists locate an index fossil in the rock layer they are studying, they know a lot about that rock layer and when its sediment was deposited.

79

The Fossil Record

All the fossils on Earth make up the **fossil record**. It is a valuable source of information about the history of life on Earth.

The fossil record provides evidence of the environment in which ancient organisms lived. An important part of an organism's environment is climate, or the average weather in an area.

We have learned from fossils that Earth's climates have changed over time. For example, fossils of plants that would have thrived in hot and humid areas (tropical jungle) have been found in areas that are now deserts. Such fossil evidence suggests that these areas were wetter millions of years ago than they are today.

Fossilized tree trunks can provide similar kinds of evidence. Trees do not grow as fast when the temperature is cold and dry. A petrified tree trunk with narrow growth rings suggests that the climate was probably cool and dry when the tree was living. A petrified tree trunk with wide growth rings suggests that the climate was probably warm and moist during the tree's lifetime.

This scientist is carefully brushing away the sand on this fossil.

Question for Reflection

- Sometimes scientists find fossils of ocean animals high up on a mountain. What do you think they conclude about the history of that area? Why?

Petrified wood is millions of years old and each piece is one of a kind.

80

81

Make Connections Imagine you found a fossil. What questions would you ask?

PART 2 Interpreting Rock Layers

SESSION 2

ELA CONNECTION

These suggested strategies address these English Language Arts objectives.

Explain events, procedures, ideas, or concepts in a scientific text, including what happened and why, based on specific information in the text.

Report on a topic or text in an organized manner, using appropriate facts and relevant, descriptive details to support main ideas.

Add visual displays to presentations when appropriate to enhance the development of main ideas or themes.

7 Discuss the reading

Give students a few minutes to review their questions and jot down new information. Then, have students discuss the Question for Reflection at the end of the article in small groups. Encourage students to use evidence from the text to support their ideas. Provide chart paper or white boards for students to make models to explain their thinking.

Call on Reporters to share their group's model and explanations and allow the other Reporters to respond by either agreeing or disagreeing, or offering alternative ideas.

➜ *Sometimes scientists find fossils of ocean animals high upon a mountain. What do you think they conclude about the history of that area? Why?* [When that layer of rock formed, the environment was marine. That would explain the fossils, perhaps in limestone or if the water was a shallow bay, it could be **shale**. Over time, the environment changed and was later uplifted by tectonic forces.]

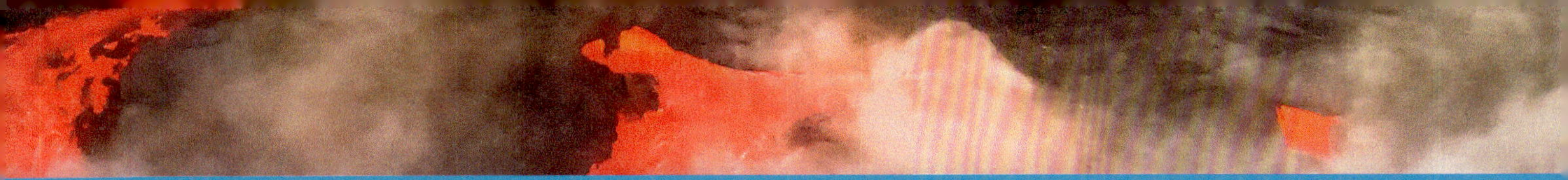

Guiding the Investigation

SESSION 3 **45 MINUTES**

8 Introduce rock layers

Distribute a copy of notebook sheet 18, *Rock-Layer Age Puzzle* to each student. The puzzle displays a model. Ask students to use the information they have gathered so far in this investigation to construct explanations to the questions about the model on their own and then discuss it with a partner.

Distribute a FOSS tray with a labeled sample of sandstone, limestone, shale, and granite to each group so they have a sample of each rock layer to observe. Give them time to write responses on the sheet working individually and then with a partner to describe information about the patterns.

Science and Engineering Practices

- Developing and using models
- Constructing explanations

Crosscutting Concepts

- Patterns

9 Review rock-forming environments

Discuss students' responses as a class and clarify any questions they have about the flat, horizontal rock layers. Students should state that rock A at the bottom is the oldest and was formed first, followed by B, C, D, and E. E is the youngest layer and was formed last. They should also know that granite is an igneous rock associated with volcanoes and that the other rocks are sedimentary.

This is a good time to confirm the types of environments that form each of the three sedimentary rocks.

- Sandstone forms from sediment deposited in sandy areas, like beaches, deserts, and dunes.
- Limestone forms from sediment deposited in water, mostly in deep ocean (marine) environments.
- Shale forms from sediment deposited in calm, muddy marine environments like swamps.

ENGLISH LANGUAGE SUPPORT
Make a chart with simple illustrations as you review the three types of environments that form the sedimentary rocks.

10 Discuss fossils in each type of rock layer

The last question might be more difficult for students. They can offer their ideas about where marine fossils and plant fossils might be found and give their reasoning for these ideas. Tell students they will be able to get more information from a multimedia activity and they can continue to add ideas to their notebook sheet response.

Have Getters return the rock samples to the materials station.

PART 2 Interpreting Rock Layers

SESSION 3

11 Engage with "Rock-Layer Movie Maker"

In "Rock-Layer Movie Maker," students can see how different environments through history resulted in different rock layers, and explore sequencing these events.

Introduce it by saying,

Each rock column will have layers of sandstone, shale, and limestone with granite on the bottom. When you look at a rock column, you will have to figure out the ancient environment that could have formed each rock layer—desert, ocean, or swamp. Click on the picture of the environment and drag it to the layer boxes (1, 2, 3, 4, or 5). Layer 1 is the oldest layer.

For each rock layer, think about the environment in which those sediments were deposited and the kinds of fossils that might be found there. And consider what effect an earthquake might have at different times.

Tips and Information will be a useful reference guide when you have questions about the rocks and fossils. At the end you can check your rock column to the actual one and make corrections if necessary.

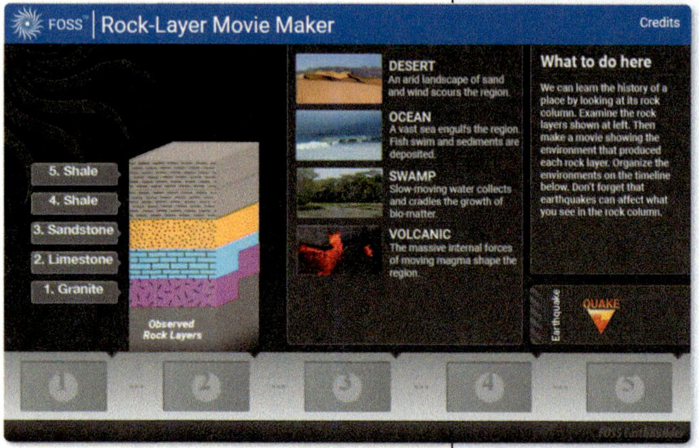

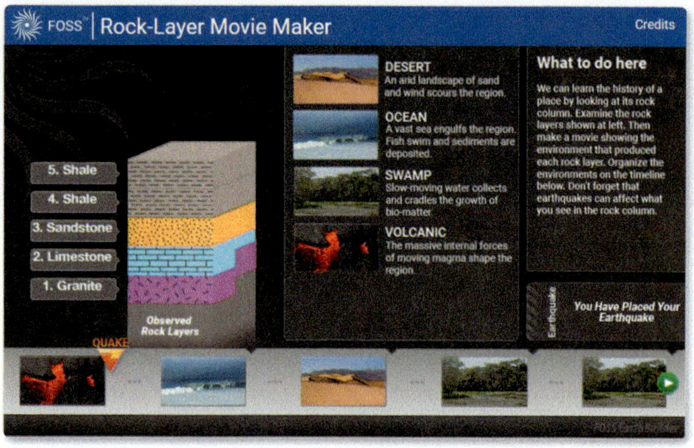

Show the tutorial by clicking Show Me How to Do It! Or you create one rock-layer movie as an example. Then have students work at computers in pairs. Have each student engage with the activity several times. Ask them to share their successes and challenges and work together to gather the information needed to match the rock layers. Ask,

➜ *Look at the rock layers in a column. Tell the story of that place by reading the rocks. What happened first, what happened next, and so on.*

Interpreting Information

SESSION 4 **45 MINUTES**

12 Read "Frenchman Mountain"

Before reading, review the rock-layer movie maker activity and share ideas or questions that came up during the experience. Then review the last question on the *Rock-Layer Age Puzzle*.

- *In what layer might you find a marine clam fossil?* [Limestone, formed in ocean (marine) environments where clams lived.]
- *In what layer might you find a plant fossil?* [Shale, formed in swamps where ferns and other plants lived.]
- *What kind of fossil might you find in the sandstone layers?* [Animal tracks formed in sand dune or on beach.]

Introduce the reading as another virtual field trip with Savannah and her family in the area outside of Las Vegas. Suggest that this reading might provide more information to add to their focus question response.

Give students a moment to preview the photographs and read the captions. You might assign different students to read aloud the parts of the narrator, Savannah, and her dad. This can be done whole-group or in groups of three. Pause to discuss the questions and prompts. Encourage students to site examples and details from the text to support their ideas.

ELA CONNECTION

These suggested strategies address these English Language Arts objectives.

Refer to details and examples in a text when explaining what the text says explicitly and when drawing inferences from the text.

Read and comprehend informational science texts.

Science and Engineering Practices

- Obtaining, evaluating, and communicating information

PART 2 Interpreting Rock Layers

SESSION 4

Frenchman Mountain

Savannah's dad said, "A friend told me about a location near Valley of Fire where deposits of silt and clay created layers of sedimentary rock called shale. Silt and clay are much smaller particles than sand. I think we should check it out this weekend. The place is called Frenchman Mountain. People visit there to search for trilobite fossils."

During the week, Savannah read information on trilobites. A trilobite is a small animal that lived on Earth's ocean floor starting about 520 million years ago. They lived in marine environments, sometimes in shallow, muddy waters. The fossils are trilobites that got trapped in sediment. They became extinct about 250 million years ago. About that same time, dinosaurs first appeared.

Savannah told her dad, "I've been looking at pictures of trilobite fossils, and there are hundreds of different ones. They all look alike, but not exactly the same."

Make Connections Have students share their ideas about why the trilobite fossils are similar but not exactly the same.

88

Trip to the Shale Quarry

On the day of their trip, Savannah gathered supplies. She asked her dad, "Will we need any tools for digging?"

He said that they could find fossils without digging. He said, "Frenchman Mountain is located on public lands managed by the Bureau of Land Management. Collecting trilobite fossils is legal and free. But if different fossils are found in the area, people must let the Bureau know."

Savannah and her dad headed out for their adventure. In less than an hour, they found the turnoff for Frenchman Mountain and started up a dirt road. The road was rough, but it brought them right to the old shale quarry. There were a couple of trucks parked there already. Savannah and her dad could see where the earlier arrivals were focusing their attention, so they walked over there.

Frenchman Mountain is about 1.2 kilometers tall.

89

Trilobite Fossils

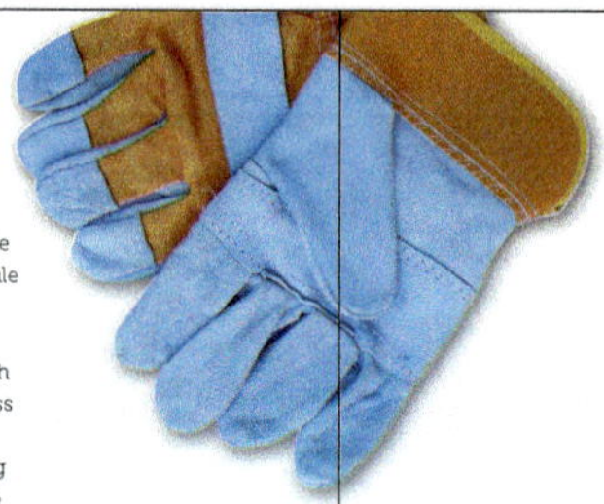

The ground was littered with hundreds of pieces of flat green shale. Savannah picked up a piece the size of a smartphone, inspected it, and said, "Look at this. I think this is part of a trilobite fossil right here." Her dad agreed, and she slipped the piece of shale into her pack.

When they got closer to the area where others were working, Savannah could see that the shale formation was actually a bunch of very thin layers. The thickest layer was only about the thickness of a slice of bread.

Savannah put on her work gloves and began eagerly inspecting the loose pieces of shale. Right away she found a treasure, a piece of shale with several complete unbroken trilobite fossils, each about 1 to 3 centimeters (cm) long.

The new find went right into her pack. Soon Savannah found another little trilobite fossil. About 10 minutes after that, Savannah suggested a break for a snack.

Back at the car Savannah said, "This green shale is full of trilobite fossils. We're far away from the ocean. How did all these ocean animals end up in rocks in this desert?"

The view from Frenchman Mountain

90

Shale in the Desert

Savannah's dad said, "All over the world scientists find rocks with marine fossils in locations far away from the ocean. Based on information you discovered about how rocks change, how do you think these trilobite fossils got here?"

Savannah thought about the sandstone strata in Valley of Fire that were tipped from horizontal layers to slanted angles. She said, "Shale forms from many layers of silt and clay. If trilobites were living in the location where the silt and clay deposited, they became trapped in the mud layers. Then when the silt and clay deposits turned to shale, the trapped trilobites turned into fossils. Later, the marine environment changed. But the trilobite fossils remained."

"The mystery is how the shale ended up forming mountains in the middle of the desert. Tectonic forces must have lifted the whole Frenchman Mountain environment. They raised the bottom of that marine sea up to this place and exposed the shale."

How did fossils of ocean animals get to desert mountains?

91

Apply Understanding Have students discuss this question. Have them think about what they have learned about the environments where sedimentary rocks form.

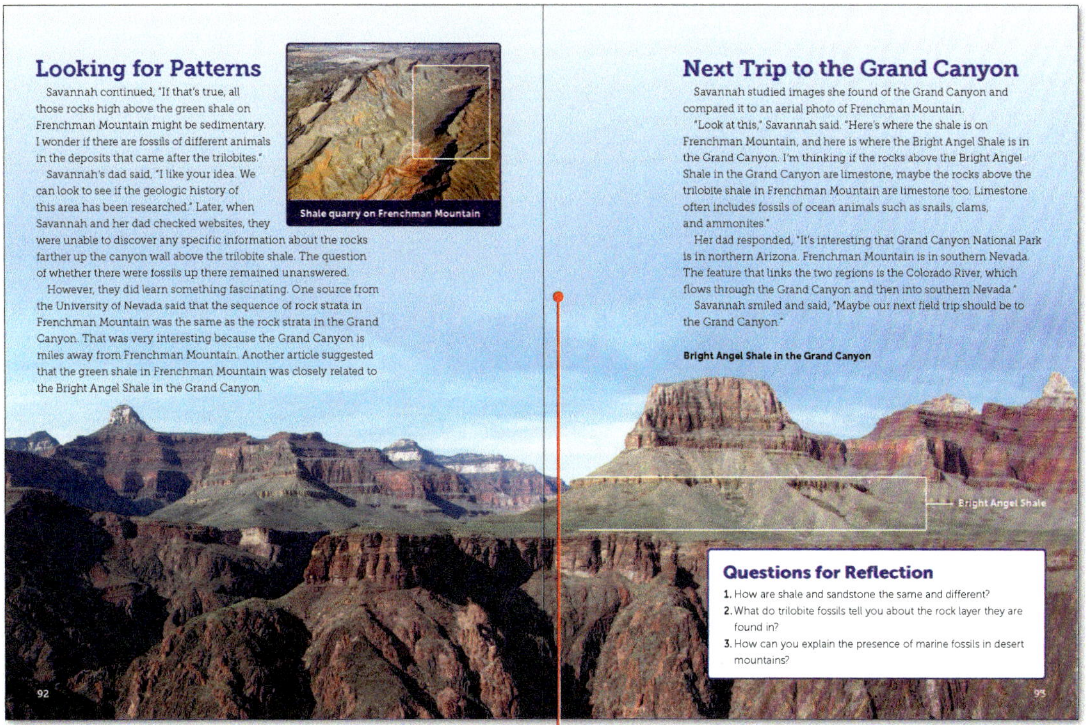

Looking for Patterns

Savannah continued, "If that's true, all those rocks high above the green shale on Frenchman Mountain might be sedimentary. I wonder if there are fossils of different animals in the deposits that came after the trilobites."

Savannah's dad said, "I like your idea. We can look to see if the geologic history of this area has been researched." Later, when Savannah and her dad checked websites, they were unable to discover any specific information about the rocks farther up the canyon wall above the trilobite shale. The question of whether there were fossils up there remained unanswered.

However, they did learn something fascinating. One source from the University of Nevada said that the sequence of rock strata in Frenchman Mountain was the same as the rock strata in the Grand Canyon. That was very interesting because the Grand Canyon is miles away from Frenchman Mountain. Another article suggested that the green shale in Frenchman Mountain was closely related to the Bright Angel Shale in the Grand Canyon.

Shale quarry on Frenchman Mountain

92

Next Trip to the Grand Canyon

Savannah studied images she found of the Grand Canyon and compared it to an aerial photo of Frenchman Mountain.

"Look at this," Savannah said. "Here's where the shale is on Frenchman Mountain, and here is where the Bright Angel Shale is in the Grand Canyon. I'm thinking if the rocks above the Bright Angel Shale in the Grand Canyon are limestone, maybe the rocks above the trilobite shale in Frenchman Mountain are limestone too. Limestone often includes fossils of ocean animals such as snails, clams, and ammonites."

Her dad responded, "It's interesting that Grand Canyon National Park is in northern Arizona. Frenchman Mountain is in southern Nevada. The feature that links the two regions is the Colorado River, which flows through the Grand Canyon and then into southern Nevada."

Savannah smiled and said, "Maybe our next field trip should be to the Grand Canyon."

Bright Angel Shale in the Grand Canyon

Questions for Reflection

1. How are shale and sandstone the same and different?
2. What do trilobite fossils tell you about the rock layer they are found in?
3. How can you explain the presence of marine fossils in desert mountains?

93

Compare and Contrast Have students compare the two photographs. What is the same and different about these landforms?

13 Discuss the reading

Have students review the Questions for Reflection.

→ *How are shale and sandstone the same and different?* [Both are sedimentary rock and form in a similar way. Sandstone forms from sand particles and shale from clay particles which are smaller than sand.]

→ *What do trilobite fossils tell you about the rock layer they are found in?* [The environment of the trilobite was a shallow marine environment, perhaps a muddy bay or swamp.]

→ *How can you explain the presence of marine fossils in desert mountains?* [The environment was swampy when the trilobites lived and died 520–250 million years ago. The environment and the landforms have changed due to tectonic forces forming mountains and by erosion and deposition.]

Crosscutting Concepts

- Stability and change

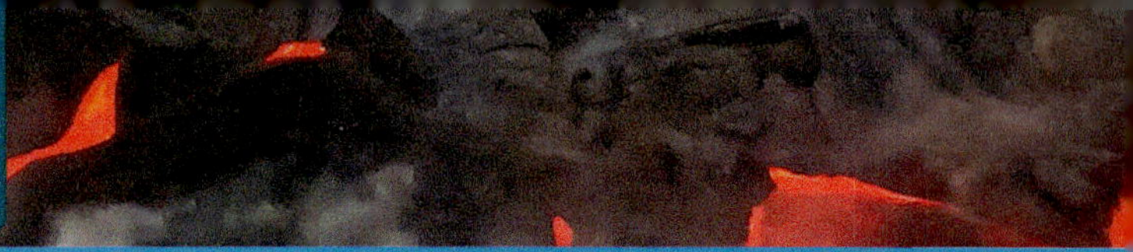

PART 2 Interpreting Rock Layers

SESSION 4

14 Finalize response to the anchor phenomenon

Ask students,

→ *What do the rock layers tell us about the age of the fossils?* [Newer layers of rocks are on top of older layers. If a fossil is in a lower layer, it is older than a fossil in a higher layer.]

→ *What do the rocks and fossils in the layers tell us about the environment at that time?* [Some rocks and fossils show the environment was aquatic or terrestrial.]

→ *What cause shifts in the rock layers?* [Earthquakes.]

15 Finalize the response to the focus/driving question

Remind students to add to their focus/driving question response.

→ *What do rock layers and fossils tell us about the history of a place?*

ENGLISH LANGUAGE SUPPORT

Encourage students to use the diagrams and charts as a reference. For students who need help getting started, provide a writing frame such as,

Fossils are found in _____. They can tell us about _____. For example, _____.

16 Assess progress: notebook entry

Have students turn in their notebooks open to the page where they've constructed explanations to the focus question, which is the same as the driving question for the phenomenon. Record your notes on a copy of *Embedded Assessment Notes.*

WHAT TO LOOK FOR

- *Sedimentary rock formations are deposited in flat layers; those layers provide information about the past environment of an area based on the kind of rocks (sandstone, limestone, or shale).*
- *Earthquakes, tectonic forces, can tilt flat, horizontal layers.*
- *Fossils found in rock layers provide evidence of organisms that lived long ago as well as clues about the age of the rock and what the environment was like when the rock layer formed.*
- *Looking at rock layer patterns provides information about how landforms are constantly changing due to weathering, erosion, deposition, canyon forming (downcutting), and earthquakes (uplifting).*

Science and Engineering Practices

- Constructing explanations

Disciplinary Core Ideas

- ESS1.C: History of planet Earth

Crosscutting Concepts

- Patterns

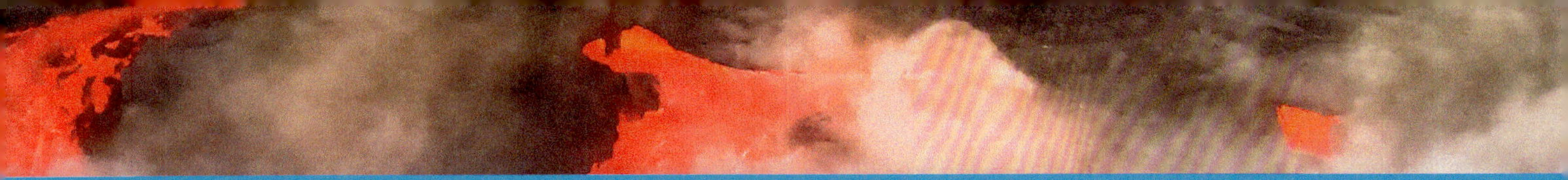

Reflecting on a Phenomenon

17 Review three-dimensions for Investigation 4

Distribute one or two self-stick notes to each student. Ask students to cut each note into three pieces. Have them take 3 minutes to look back through their notebook entries to find the three most important things they learned in Investigation 4. Students should include at least one science and engineering practice, one disciplinary core idea, and one crosscutting concept. They should tag those pages using the three self-stick notes.

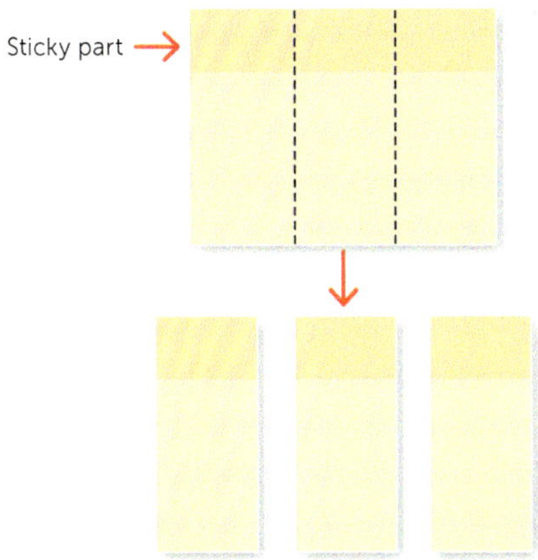

When students have completed their notebook review and placed their tags, lead a short class discussion to create a list of three-dimensional statements that summarize what students have learned in this investigation.

- We mapped the locations of significant volcanoes and earthquakes using longitude and latitude data and analyzed and interpreted the data to reveal patterns: most earthquakes and volcanoes occur near the edges of continents that correspond with the boundaries of tectonic plates.
- We gathered information to communicate the process of sedimentary rock formation. Sediments are deposited in horizontal layers, with the oldest layers on the bottom and the newest layers on top; the size of particle and type of environment determine the kind of rock that forms (beach or dune for sandstone, swamp for shale, and ocean for limestone). We used a modeling tool to analyze and interpret data about rock layer formations over time.
- We gathered information about how different kinds of fossils provide evidence of organisms that lived in past environments; rocks layers tell the story of how environments change over time.

18 Review the module

Have students work with a partner and review the reading "Landforms on Display." Have them look at the photographs, select several images in each category, and respond to the question "What happened here?" Students should explain what they know about how each landform was created by weathering and erosion, deposition, eruptions, or crust movements. Encourage students to make models in their notebooks to use as they explain their thinking.

THREE-DIMENSIONAL LEARNING OBJECTIVES

- **Students analyze and interpret data sets of active volcanoes and large earthquakes by mapping locations on a map to reveal patterns. (Part 1)**
- **Students obtain information from media to construct explanations about the geological history revealed by patterns of rock layers and the presence of types of fossils. (Part 2)**

PART 2 Interpreting Rock Layers

SESSION 5 45 MINUTES

19 Assess progress: *Posttest*

Give the *Posttest* at least one day after the wrap-up review. Distribute a copy of the *Posttest* to each student or schedule it on FOSSmap. You can read the items aloud, but students should respond to the items independently.

Collect the *Posttest*, code the responses or review the responses on FOSSmap, and plan for a self-assessment session with students. More information about self-assessment activities can be found in the Assessment chapter. The latest Assessment Coding Guides are found on FOSSweb and are used on FOSSmap.

Differentiation Opportunities

Below is a sample of student-choice activities related to this investigation for group or individual projects. Students can select an activity that is of special interest to them, or the class can choose a project for collaboration. Extensions and project ideas are written for students and accessible on FOSSweb.

Language Extension

- Research your state fossil

Science Extensions

- Explore your local region for evidence of rapid changes in landforms
- Review the resources and information on the USGS volcano hazard site
- View the interviews with USGS scientists in the video *Remembering Ridgecrest*

Environmental Literacy Extensions

- Check with your city or state for natural hazards and the recommended emergency preparedness plans and procedures
- Plan and take a field trip to a local geological site or roadcut

Home/School Connection

Students gather recent news reports about landforms and the natural or human-caused processes that cause change to landforms—storms and floods, landslides, earthquakes, and severe wind.

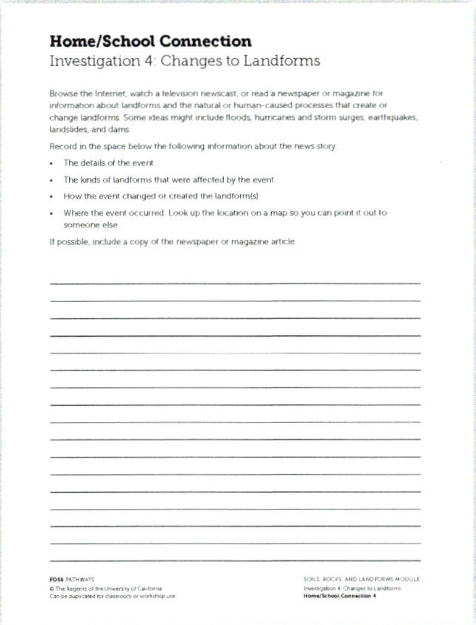

Home/School Connection

Investigation 4: Changes to Landforms

Browse the Internet, watch a television newscast, or read a newspaper or magazine for information about landforms and the natural or human-caused processes that create or change landforms. Some ideas might include floods, hurricanes and storm surges, earthquakes, landslides, and dams.

Record in the space below the following information about the news story:

- The details of the event.
- The kinds of landforms that were affected by the event.
- How the event changed or created the landform(s).
- Where the event occurred. Look up the location on a map so you can point it out to someone else.

If possible, include a copy of the newspaper or magazine article.

FOSS PATHWAYS
© The Regents of the University of California
Can be duplicated for classroom or workshop use.

SOILS, ROCKS, AND LANDFORMS MODULE
Investigation 4: Changes to Landforms
Home/School Connection 4

No. 4—Home/School Connection

ASSESSMENT

Soils, Rocks, and Landforms

Assessment

Soils, Rocks, and Landforms

The FOSS Assessment System for Grades 3–5

"Assessment is like science. ...To assess our students, we plan and conduct investigations about student learning and then analyze and interpret data to develop models of what students are thinking. These models allow us to predict the effect of additional teaching, addressing the patterns we notice in student understanding and misunderstanding. Assessment allows us to improve our teaching practice over time, spiraling upward" (*2016 Science Framework for California Public Schools, Kindergarten through Grade 12,* chapter 9, page 3).

Assessments should be designed to meet specific purposes. One size does not fit all. The FOSS assessment system provides ample opportunities for both formative and summative assessment. Formative assessments provide short-term information about learning by making students' thinking visible in order to guide instructional decisions. Summative assessments provide valid, reliable, and fair measures of students' progress over a longer period of time, such as at the end of a module, or the end of the year. The purpose for the assessment determines the choice of instruments that you will use.

The FOSS assessment system is designed to assess students in cycles: short, medium, and long. The assessment tasks allow students to demonstrate their facility with three-dimensional understanding of science.

Short cycle. Embedded assessment opportunities are incorporated into each part of every investigation. These assessments use student-generated artifacts, including science notebook entries, responses to focus questions, response sheets, and **performance assessments**. Embedded assessments provide daily monitoring of students' learning and practices to help you make decisions about instructional next steps. Embedded assessments using science notebooks provide evidence of students' conceptual development. Performance assessments focus on science and engineering practices and crosscutting concepts, as well as disciplinary core ideas.

CONTENTS

NOTE

FOSS assessments are reviewed and updated annually. For the most up-to-date assessment masters, answer sheets, and coding guides, go to FOSSweb.

The FOSS Assessment System for Grades 3–5

I-Check (short for "I check my own understanding") opportunities occur at the end of every 1–2 investigations. These assessments are actually hybrid tools that provide summative information about students' achievement, and they are even more powerful when used for formative assessment. Daily embedded assessments provide a quick snapshot of students' immediate learning, and I-Checks challenge students to put this learning into action in a broader context. Students must think about the science and engineering practices, disciplinary core ideas, and crosscutting concepts they have been learning, and know when, where, and how to use them. I-Checks also provide opportunities for guided self-assessment, an important skill for future learning and the development of a growth mindset. Timely, meaningful feedback can help a student focus attention on areas that need strengthening. When a student responds to feedback, you can develop an even more precise understanding of the student's learning. A feedback/response dialogue can develop into a highly differentiated path of instruction tailored to the learning requirements of individual students.

Medium and long cycle. *Survey/Posttest,* interim benchmark assessments, and portfolios are tools provided for medium- and long-cycle assessment. Students take the *Survey* before instruction begins. This entry-level assessment provides you with information about students' prior knowledge and developing practices. It indicates modifications that might be needed for specific students or groups of students.

The *Posttest* is given at the end of a module. It provides summative information about students' three-dimensional learning. It also lets students compare their *Survey* responses to those on the *Posttest* to see how their understanding has grown. You can also use the *Posttest* for formative instructional evaluation to note the things you might want to focus on or do differently next time you teach the module.

Interim benchmarks are offered as an optional tool and can be given during a module (when you want information about how students are progressing in regards to a performance expectation), at the end of the module, or at the end of the school year. These assessments specifically address NGSS performance expectations and require students to apply what they have learned to new contexts. The interim assessment tasks can also inform instruction when used formatively to plan for next year's instruction.

Students can also collect work samples in a portfolio as they work through the module. At the end of each investigation, they can create summative responses to document their three-dimensional learning.

Assessment for the NGSS

A Framework for K–12 Science Education (National Research Council, 2012), *Next Generation Science Standards* (National Academies Press, 2013), and the *Developing Assessment for the Next Generation Science Standards* (National Research Council, 2014), and many state frameworks provide a new vision for science education. These documents emphasize the idea that science education should resemble the way that scientists work. Students plan and conduct investigations, gather data to construct explanations, and engage in argumentation to build their understanding of the natural world. They apply that knowledge to engineering problems and design solutions. Students are expected to construct and discuss explanations and model systems in increasingly more sophisticated ways as they move through the grades. Assessment plays an important role in this new vision of science education—assessment is the bridge between teaching and learning.

"Assessment plays an important role in this new vision of science education—assessment is the bridge between teaching and learning."

Assessment tasks should consist of multiple components in order to measure all three dimensions of science and engineering learning. The FOSS assessment system provides multiple tools and strategies to assess the three dimensions: (1) science and engineering practices, (2) disciplinary core ideas, and (3) crosscutting concepts. These tools and strategies provide evidence about what students can do, their developing conceptual understandings, and the connections that they are making among disciplines. Entry-level assessments, given before instruction begins, show what students can do and what they know before they begin a new module.

Assessment systems should include formative and summative tasks. Formative assessment tasks are embedded in the curriculum at key stages in instruction. These tasks are designed to support you in collecting and analyzing data about students' conceptual understanding. Notebook entries, responses to focus questions, response sheets, performance assessments, and oral presentations/interviews provide the information you need to decide what students should do next to move toward a learning goal. FOSS suggestions for next-step strategies help you address students' developing conceptions and provide information for differentiated instruction.

Assessment for the NGSS

Summative assessments are designed to provide valid, reliable, and fair measures of students' progress. The FOSS system includes three types of summative assessments: *Posttests,* interim assessments, and portfolios. These assessments include open-ended constructed-response items as well as some multiple-choice and short-answer items. *Posttests* are given at the end of a module, and interim assessments can be given twice during a module or as an end-of-year assessment. Students can also collect work products at the end of each investigation for inclusion in a portfolio. Coding guides found on FOSSweb provide you with guidance when evaluating these assessments. All written assessments are consistent with grade-level writing and mathematics for ELA/Literacy and for Mathematics.

The main purpose of the FOSS assessment system is to support classroom instruction—to provide the bridge between teaching and learning. You need information daily about what students have learned or may be confused about. FOSS has developed a technique in which you spend only 10 minutes after a lesson, using a reflective-assessment practice (explained in detail later in this chapter), to gather data to determine instructional next steps. *Are students ready to move on to the next lesson, or do they need some additional clarification?* FOSS research has shown that the reflective-assessment practice provides evidence-based information that is crucial for differentiating instruction for all students. This practice can make a significant difference in students' overall achievement.

"FOSS research has shown that the reflective-assessment practice provides evidence-based information that is crucial for differentiating instruction for all students. This practice can make a significant difference in students' overall achievement."

The FOSS assessment system is based on research that describes a conceptual framework and learning objectives to provide evidence of students' progressive learning (see the Framework and NGSS chapter). These assessment tasks provide valid, reliable, and fair evidence of students' learning. (See the Benchmark Assessment section for a more detailed description of the design behind the FOSS assessment system.)

Assessment systems should include an interpretive system and locate students along a sequence of progressively more complex understanding. FOSS provides extensive support for interpreting assessment information. For each embedded assessment, specific information in the Getting Ready and Guiding the Investigation sections describes what students do and what to look for in student responses to assess progress. Coding guides, found on FOSSweb, are provided for each item on benchmark assessments (I-Checks and the *Survey/Posttest*, as well as interim assessments). Samples of student work, especially for open-response questions, are also available on FOSSweb. Resources in this chapter and on FOSSweb provide you with information about how to use the coding guides, as well as what to do for next steps when students need to spend more time on an idea, or to look at it from a different perspective to see more connections.

"Assessment is no longer a stress factor for students or teachers. It encourages all to adopt a growth mindset—if I know where my strengths and weaknesses are and I continue to be thoughtful and work hard, I can make progress."

You can use FOSSmap to have students take assessments online and generate a number of diagnostic and summary reports (delivered as PDFs). Some of these reports provide information about class progress; others provide individual students and parents with information about what students know and what they still need to work on. (See the FOSSmap and Online Assessment section in this chapter.)

The FOSS assessment system was developed over a period of 5 years with data from more than 500 teachers and their students. As a result, you can employ this assessment system for the benefit of your students, knowing that your students will achieve more. Perhaps even more important is the change in classroom culture that occurs when assessment is thoughtfully employed as the bridge between teaching and learning. Assessment is no longer a stress factor for students or teachers. It encourages both you and your students to adopt a growth mindset. The growth mindset says that if I know where my strengths and weaknesses are and I continue to be thoughtful and work hard, I can make progress. It models what scientists do. Scientists use the information they have to argue for the best explanation, but they keep an open mind, so when new evidence emerges, they can incorporate that into their thinking too. That's also what good curriculum and assessment are all about.

Embedded Assessment

Assessment is the bridge that connects teaching and learning. Assessing students on a regular basis gives you valuable information that guides instruction and keeps families informed about students' progress.

Embedded assessments are suggested for most investigation parts. You will find a statement of what and when to assess in the Getting Ready section of each part of each investigation. For example, here is the **Getting Ready** Plan for Assessment step for Part 1 of the first investigation.

Plan for Assessment

- Review Coding Guide for *Survey* (identical to *Posttest*) F
- Review Step 19, "What to Look For" in the notebook entry F

At the point in the lesson when students create a work product to be assessed, you will find a step in **Guiding the Investigation**. It provides a bulleted list of what to look for when you review the students' work for disciplinary core ideas, science and engineering practices, and crosscutting concepts.

Science and Engineering Practices
- Analyzing and interpreting data

Disciplinary Core Ideas
- ESS2.A: Earth materials and systems
- ESS2.E: Biogeology

Crosscutting Concepts
- Patterns

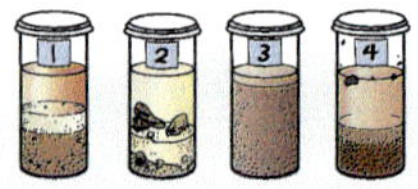

19 Assess progress: notebook entry

Have students hand in their notebooks open to the page on which they answered the focus question. You might also want to take a quick look at students' preliminary claims to help you prepare for the next set of lessons.

WHAT TO LOOK FOR

- *Based on analysis and interpretation of firsthand observations, students write that soil is composed of different sizes of rock (e.g., sand, gravel, pebbles) and humus (decaying plants and animals).*
- *The components of the soil form layers (patterns) when each sample is mixed with water and allowed to settle. The amount of each component provides some evidence for where the soil came from. The soil with the biggest rocks might be from the mountains. The soil with the most humus might be from the forest.*

20 Save one set of soil samples

Save one group's set of vials as a reference when they refine their claims in Investigation 2, Part 2, Step 17.

Have students return all equipment to the materials station. Clean out the rest of the vials. Set up two basins, one empty and the other filled with water. Students can dump the soil and water into the first basin and rinse the vials in the second basin. All of the soil and water should be recycled appropriately outdoors.

46 **FOSS** PATHWAYS

Performance Assessments

Assessing the three dimensions envisioned in the NRC *Framework* and the NGSS performance expectations challenges students to be engaged in science and engineering practices in order to build disciplinary core ideas bridged by crosscutting concepts. This is an everyday occurrence in the FOSS curriculum. One part in each investigation has been designated as a performance assessment for you to formatively check students' progress for all three dimensions. You peek over students' shoulders while they are in the act of doing science or engineering. You take note of what they are doing and discussing and sometimes conduct short interviews. Observing the rich conversation among students and the actions they are taking to investigate phenomena or design solutions to problems provides important information about student progress. At times, you might step in with a 30-second interview to ask a few carefully crafted questions to learn more about students' deeper conceptual understanding and practices. The What to Look For bullets in each performance assessment step will help you focus on pertinent science and engineering practices, disciplinary core ideas, and crosscutting concepts. You can record student progress on the *Performance Assessment Checklist* for each part.

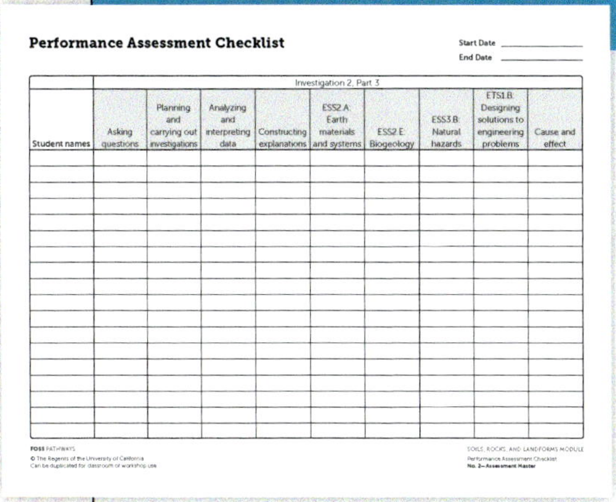

Performance Assessment Checklist

Start Date ____________
End Date ____________

	Investigation 2, Part 3								
Student names	Asking questions	Planning and carrying out investigations	Analyzing and interpreting data	Constructing explanations	ESS2.A: Earth materials and systems	ESS2.E: Biogeology	ESS3.B: Natural hazards	ETS1.B: Designing solutions to engineering problems	Cause and effect

FOSS PATHWAYS
© The Regents of the University of California
Can be duplicated for classroom or workshop use

SOILS, ROCKS, AND LANDFORMS MODULE
Performance Assessment Checklist
No. 2—Assessment Master

Science Notebook Entries

Making good observations and using them to develop explanations about the natural world is the essence of science. This process calls for critical thinking and communication skills. Science notebook entries are designed specifically to help you understand the practices, crosscutting concepts, and scientific explanations and knowledge that students are developing.

Notebook entries serve as assessments for learning. Each part of each investigation is driven by a **focus question**. Each part usually concludes with students writing a response to the focus question in their notebooks. Their responses reveal how well they have made sense of the investigation. Prepared **notebook sheets** or **free-form notebook entries** provide information about how students make and organize observations and their thought process when analyzing and interpreting data. Finally, **response sheets** provide more formal embedded-assessment data in each investigation. These assessments focus on specific scientific knowledge, practices, or crosscutting concepts that students often struggle with, giving you an additional opportunity to help students untangle concepts.

NOTE

You only need 10 minutes after a lesson to review student work and gather evidence of learning. See the reflective-assessment practice on the next page.

Embedded Assessment

Using the Reflective-Assessment Practice

Continuous formative assessment is a teaching practice that supports student progress. One of the keys to formative assessment is frequency. The more often you can gather evidence about students' progress, the more able you will be to guide each student's path to understanding. The **reflective-assessment practice** is a proven method for gathering that information. It takes little time and has a big impact on students' learning.

Reflective-Assessment Practice

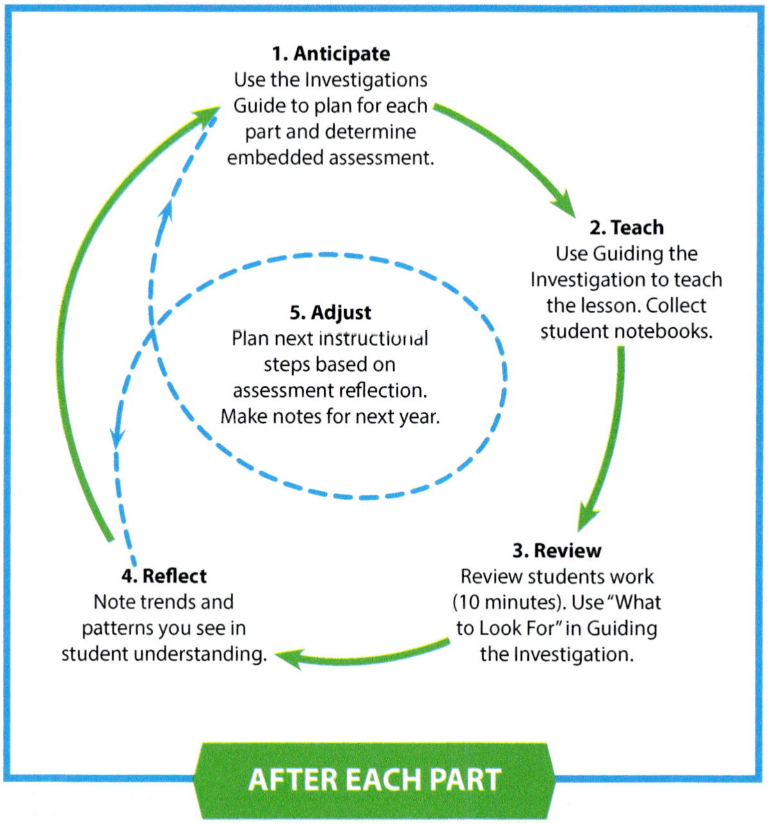

Print or make copies of ***Embedded Assessment Notes*** to record students' progress with the embedded assessments in each part of an investigation (see the What to Look For bullets). If you don't think you can review every student's work in 10 minutes, choose a random sample. Many teachers are pleasantly surprised to discover how many students they can review in that short amount of time. The important thing is that you are looking at student work on a regular basis as often as possible and looking for patterns that reveal strengths and areas that need additional support.

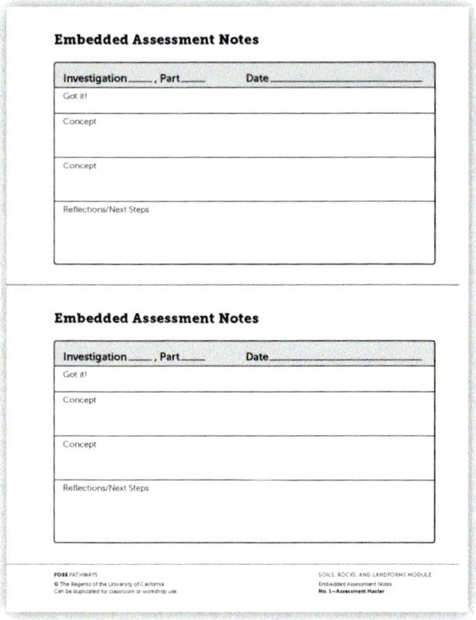

Embedded Assessment Notes

Investigation ___, Part ___ Date ___

Got it!

Concept

Concept

Reflections/Next Steps

Embedded Assessment Notes

Investigation ___, Part ___ Date ___

Got it!

Concept

Concept

Reflections/Next Steps

FOSS PATHWAYS
© The Regents of the University of California
Can be duplicated for classroom or workshop use.

SOILS, ROCKS, AND LANDFORMS MODULE
Embedded Assessment Notes
No. 1—Assessment Master

You can record notes for two lessons on each page.

1 Anticipate. Check the Getting Ready section for the suggested embedded assessment for the part you are planning. Before class, fill in the investigation and part number along with the date on *Embedded Assessment Notes*. Check the assessment step in Guiding the Investigation, and fill in the things you are looking for in student thinking. Limit your review to one or two important ideas.

2 Teach. Follow the steps in Guiding the Investigation.

3 Review. Collect the notebooks at the end of class. Have students turn in their notebooks *open to the page you will be reviewing*. (This may sound trivial, but it will save you a lot of time.) If you didn't write in your What to Look For before teaching the lesson, do that now by checking the assessment step in Guiding the Investigation. Use *Embedded Assessment Notes* to record what you observe. Make a tally mark for each student who "got it"; write in names and notes for students who need help. Spend no more than 10 minutes on this review.

NOTE

We encourage you to stop after 10 minutes even though you may be tempted to spend more time. If you keep this process quick and easy, you are more likely to use it frequently. Frequency is key to the success of this practice.

Embedded Assessment Notes

Investigation 1, Part 1	Date 3/16/23
Got it!	~~||||~~ ~~||||~~ ~~||||~~ ~~||||~~
Concept	Soil is composed of different sizes of rock (e.g., sand, gravel, pebbles) and humus. Angelina, Kai, and Robert did not mention humus in their soil descriptions.
Concept	The amount of each component provides evidence for where the soil came from. Ellie, Ming, and Sam did not state any evidence for location.
Reflections/Next Steps	Most students describe soil components accurately and understand that the amounts of the components provide clues to where the soil came from. Have Angelina, Kai, and Robert return to their soil drawings and compare to the vial samples. Have Ellie, Ming, and Sam repeat the shake test.

Note what to look for and make notes as you review student work.

Write reflections and next steps here after 10 minutes of review.

4 Reflect. Take 5 minutes to summarize the trends and patterns (highlights and challenges) you saw, and record notes in the "Reflections/Next Steps" section.

5 Adjust. In the same section of the sheet, describe the next steps you will take to clarify any problems, or note highlights you saw in students' progress. Some suggestions appear later in this chapter and accompany the coding guides. This is the defining factor in formative assessment. You must take some action to help students improve. If you do this process frequently, the next steps required should take only a few minutes of class time when the next part begins.

Benchmark Assessment

The FOSS benchmark assessments are carefully designed to help you look at students' progress across the module. Each of these assessments includes items that incorporate the three dimensions described in the *NRC Framework:* science and engineering practices, disciplinary core ideas, and crosscutting concepts. Benchmark assessments provide a broader focus on students' learning than do embedded assessments. Benchmark assessments require students to determine when and how pieces of knowledge, and larger conceptual relationships, need to be recalled or applied.

Survey

The *Survey* is administered a few days before instruction begins. Students are often uneasy about having to take a "test" when they haven't yet had the instruction they need in order to do well. It is important for them to know that the *Survey* will help you determine what they already know and what they need to learn, so that you can plan extra support when needed. Stress to your students that they will not be graded on this assessment. Help students view the *Survey*, as well as all of the assessments, as learning tools. At the end of the module, have students compare their answers from the *Survey* to their answers on the *Posttest*. This gives students the opportunity to see how their knowledge changed.

NOTE
Alternatively, students can take the *Survey*, I-Checks, and *Posttest* online. Most of the items will be automatically coded for you. You can print students' answers from the *Survey* for comparison with the *Posttest*. See the FOSSmap and Online Assessment section in this chapter.

I-Checks

At the end of every 1–2 investigations, students take an I-Check benchmark assessment. I-Checks (short for "I check my own understanding") provide students with an opportunity to demonstrate their learning. When you return the I-Checks, students have another opportunity to reflect on their understanding through self-assessment activities.

If students complete an I-Check using paper and pencil, determine their progress by reviewing each item using the coding guides provided on FOSSweb. When you code the open-response items, we recommend that you code them one item at a time; for example, code item 8 for all students, then move on to item 9, then 10, and so on. Even though you have to shuffle papers more, you will find that it takes less time overall to code the assessments. Evaluating one item at a time, rather than one student at a time, allows you to establish a mindset for each item, think about the whole class's performance on that item, and plan the next steps you might take. Another benefit to following this procedure is that you will code more consistently.

I-Checks are most valuable when you use them as formative assessments. Studies have shown that students learn more when they take part in evaluating their own responses. To do this, code students' tests, but do not put any marks on them. Thinking about the content of the questions comes to a halt when students see marks on their papers (especially grades) made by their teachers. Using the self-assessment strategies described later in this chapter is an important step in developing a class culture focused on a growth mindset that values assessment as a learning process.

Posttest

Students take the *Posttest* after all the investigations are completed. It can be administered in any of the ways described for the other benchmark assessments. After coding the *Posttests*, return them and the *Surveys* to students. Have students compare their *Survey* and *Posttest* responses. Discuss the changes that have occurred.

Use the *Posttests* for formative evaluation of the module. Make notes about things you might want to focus on or do differently the next time you teach the module.

When Grades Are Required

"If we are always focusing on formative assessment, how do we give grades?" is a frequent question. Research shows that grades often inhibit, rather than motivate, effort and learning. To support a growth mindset, FOSS recommends that for the purpose of grading, you ask students to create revised summary responses after reflecting on their work. For example, if you need to grade a notebook entry, have students revise that entry after discussion on a separate sheet of paper and turn it in for the grade. An important part of the student's learning process is that they may work with other students or use a class discussion to improve their entry before submitting it.

When developing a grade from I-Checks, it's best not to calculate a straight percentage. Instead, you could choose three or four items that students have the opportunity to reflect on through self-assessment strategies. Ask students to rewrite these items and turn the revised responses in for a grade. When you follow this process, students are always clear about when they will be graded and know that they always have the opportunity to make their work better, which helps develop a growth mindset.

If you must base benchmark assessment grades on a percentage, remember that codes represent levels of mastery, and are *not the same as points*. One commonly used practice is that you add the codes and determine a percentage (or let FOSSmap calculate this percentage for you) but use non-traditional cutoffs for your grades. Due to the conceptual design of FOSS benchmark assessments, we suggest that 80% should be an A, rather than the traditional 90%, as a starting point for your benchmark grading system.

NOTE

Recognition, **Conceptual**, and **Strategic** levels represent students performing at or above grade level expectations.

Notions indicates students that have not made progress.

Benchmark Assessment

Portfolios

If you are using portfolios, give students a *Portfolio Checklist* and a folder to hold the checklist and their work samples. At the end of each investigation, suggest an item or two that they might add to the portfolio to demonstrate their three-dimensional learning. Students can photograph and print notebook pages or create a derivative product from the notebook page to be included in the portfolio. Notebooks should never be graded; they should be a risk-free environment for students to record their thinking. If students are going to include something from their notebook, they should have a chance to improve it before adding it to the portfolio. A sample *Portfolio Checklist* can be found with the other FOSS assessment charts.

Interim Assessments

These assessments directly target specific NGSS performance expectations. They are not meant to be diagnostic for daily instruction using FOSS curriculum, nor are they meant to be graded. Rather, they are generic tasks that students should be able to answer given any curriculum used to teach the NGSS. They are a good tool for students to use to practice for state or district tests. Interim assessments can be given after students complete an investigation focusing on a particular performance expectation, or you can give them at the end of the year for program evaluation, especially when a state or district test is not given at your grade level. Interim assessments provide another opportunity for students to practice and assess their ability to apply concepts to unfamiliar contexts.

These tasks begin with a scenario that students read to set the context for the practices and items that will be involved in the task. You can opt in many cases to include a hands-on experience. Each task then consists of a number of related items that students answer. Most are constructed-response items, but a few may be multiple-choice, multiple-answer, or short answer.

For fourth graders, use these items as a teaching and learning experience. That means that you support students by reading the text aloud, and have the class work through the task together. You provide any support the students need, short of giving them the answers. After students complete the tests, use them formatively, like the I-Checks, to evaluate students' progress. Keep in mind that the purpose of these assessments in fourth grade is to help your students develop a growth mindset.

Benchmark Assessment Design

The foundation of the FOSS assessment system rests on three pillars: a conceptual framework, a student-progress model, and the NGSS performance expectations. The **conceptual framework** for this module can be found in the Framework and NGSS chapter. It is a segment from a larger learning progression that FOSS has carefully engineered for grades K–8. The concepts you see in the framework span grades K–8, and the bullets designate learning that is important for fourth graders. The **NGSS performance expectations** also guide assessment design, providing a sampling of end goals and highlighting that science is a three-dimensional enterprise that includes science and engineering practices, disciplinary core ideas, and crosscutting concepts. The **student-progress model** is displayed in the table below. The numbers in the table indicate the *code* or *level* of student learning, which is not the same as *points* assigned on more traditional tests. It is important that these levels be treated as conceptual indicators and not be assigned as grades.

Level	Description
Strategic (code 4)	Performance on an item or assessment shows exceptional understanding of three-dimensional learning. Students are able to apply their knowledge of practices, disciplinary core ideas, and crosscutting concepts to explain novel phenomena or solve new problems. Students at this level continue to build a network of knowledge, practice, and crosscutting concepts to bridge disciplines. They can apply all of those to real-world phenomena and design problems.
Conceptual (code 3)	Performance on an item or assessment shows well-developed understanding of three-dimensional learning. Students are making connections among practices, core ideas, and crosscutting concepts in order to answer more complex questions and solve more complex problems than at the recognition level. To get to the next level (strategic), students need to continue to build connections and be able to transfer this knowledge from the classroom to real-world phenomena.
Recognition (code 2)	Performance on an item or assessment shows developing understanding of three-dimensional learning. Students have built a foundational repertoire of pieces of knowledge and practices and use academic language with greater facility. To get to the next level (conceptual), students need to continue to add knowledge about practices, core ideas, and crosscutting concepts, and then build connections among those pieces to form more complex understandings about phenomena.
Notions (code 1)	Performance on an item or assessment shows relatively little understanding of three-dimensional learning. Students may include some scientific vocabulary or recall of simple facts or procedures, but there is little evidence of the impact of instruction. To get to the next level (recognition), students need to develop practices, begin to incorporate academic language in their communication, and construct more pieces of knowledge that help explain phenomena.

Recognition, **Conceptual**, and **Strategic** levels represent students performing at or above grade level expectations. **Notions** indicates students that have not made progress.

Benchmark Assessment

Through FOSS research (ASK Project, 2003–2009) and that of others (most notably, Dylan Wiliam and Carol Dweck), FOSS developers have concluded that the most important assessment goal is to support learning in a way that helps students develop a growth mindset. That means that the focus must be on formative assessment.

It also means that we don't use traditional scoring. Instead, you code student responses to categorize the level at which students have demonstrated competence. Remember that a code 1 does not indicate one *point*. In fact, their work is in the notions level (code 1) category. They have attempted to answer the question, but the answer shows little influence of instruction. You will also notice that some items can be coded no higher than a 2 (recognition level), even though the student-progress model has four levels. That particular item asks students only to demonstrate that they can perform a simple practice or have a piece of knowledge (a recognition-level proficiency) essential to a broader understanding. When it is important to know what pieces students have or don't have in order to determine how to help them improve, a simpler question can help identify differentiated instruction that is necessary for some students. These less-complex questions are important in terms of providing diagnostic formative information.

Coding Benchmark Assessments

Answer sheets and coding guides for all benchmark assessments are available on FOSSweb. A two-page spread has been dedicated to each page of each assessment. On the next page, we show an example for coding a multiple-answer and an open-response question.

NOTE
Coding guides for interim assessments are found on FOSSweb.

Recording Benchmark Data

Use ***Assessment Record*** to record students' responses on the benchmark assessments (*Survey*, I-Checks, and *Posttest*). Follow this procedure if you plan to use the I-Checks for formative assessment.

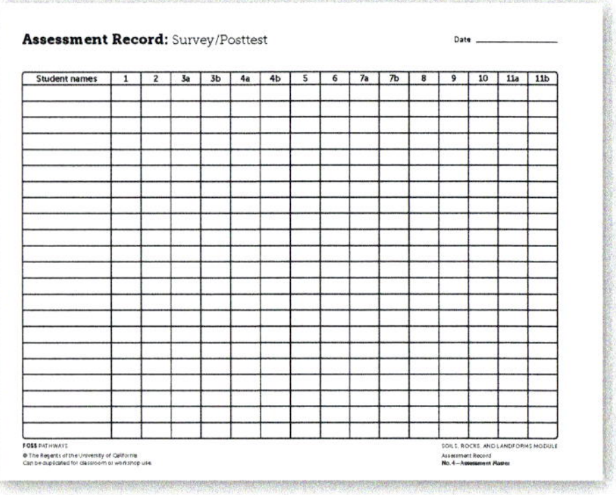
Assessment Record: Survey/Posttest Date ________

Student names	1	2	3a	3b	4a	4b	5	6	7a	7b	8	9	10	11a	11b

1. Record students' responses on *Assessment Record* or on a spreadsheet. For multiple-choice and multiple-answer items, record the letter of the response rather than the code. (Later, if you need a total score, you can replace the letter responses with numbers.) For multiple-answer items, check the coding guide; for short-answer questions, record the word or the code, depending on the answer patterns you want to see. Do not make any marks on students' papers. (Students will do that later.)

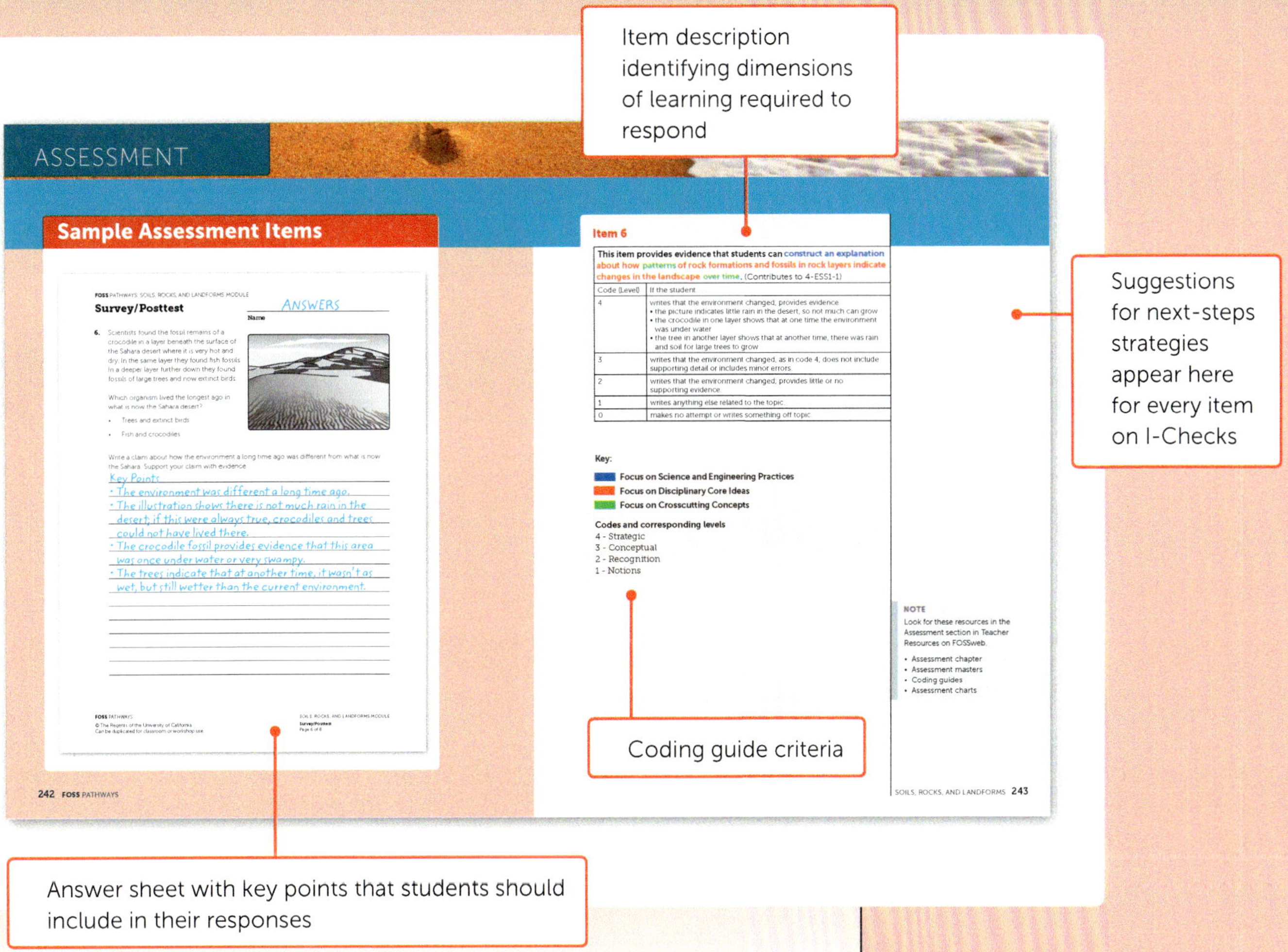

2. Look for patterns of right and wrong answers, so that you can determine which questions will need follow-up reflection and discussion by students.
3. Review students' open-response answers. Record codes on *Assessment Record*. Note concepts or ideas you want to bring back for students' reflection in self-assessment activities.
4. Choose three or four items from the assessment that you want to discuss with students during class time. Determine the next steps you want to introduce. (See the following section for more information about next-step strategies.)

NOTE

If students take the benchmark assessments online, FOSSmap does most of the coding for you. You will only need to code open-response items and check codes for short-answer items. Then you can run a variety of class and individual diagnostic reports.

Next-Step Strategies

The Assessing Science Knowledge (ASK) Project was funded by the National Science Foundation in 2003. For 6 years, the FOSS development team worked with more than 500 teachers and their students as research partners. The evidence provided by these studies proved that assessment is only worth doing if follow-up action is taken to improve understanding. Self-assessment provides students the opportunity to be responsible for their own learning and is a very effective tool for building students' scientific knowledge and practice.

Self-assessment is more than reading correct answers to the class and having students mark whether they got the right answer. Self-assessment must provide an opportunity for students to reflect on their current thinking and judge whether that thinking needs to change. This kind of reflective process also helps students develop a better understanding of what is expected in terms of well-constructed responses.

Self-assessment requires deep, thoughtful engagement with complex ideas. It involves students in whole-class or small-group discussion, followed by critical analysis of their own work. Therefore, focus on three or four questions from an I-Check, rather than on the entire assessment. The techniques described here are meant to give you a few strategies for entering the process of self-assessment. There is no single right way to engage students in this process, but it works best when you change the process from time to time to keep it fresh. The strategies listed here are sorted into two groups: (1) strategies for whole-class feedback and (2) strategies for individual-student and small-group feedback.

NOTE
If students have taken an assessment online, print out their responses (the Student Responses Report) and use a projection system and the assessment masters or the PDF to show students the items under discussion.

Strategies for the Whole Class

Multiple-choice discussions. Students sit in groups of three or four, depending on how many possible answers there were for a given question. You assign an answer to each student (not necessarily the answer he or she chose). Each student is responsible for explaining to the group whether the assigned answer is correct and why or why not.

Multiple-choice corners. When the class is equally split on what students have chosen as the correct response, or only a few students got the correct answer, have them meet in different corners of the room. Those who chose A go to one corner, those who chose B go to another corner, and so on. Each corner group needs to come up with an argument to convince the other corners that their answer is correct. As in a class debate, students are allowed to disagree with

themselves if they become convinced their position is flawed or the reasoning of another group is more convincing. They then move to that corner and continue by helping their new group shape its argument. (Don't be surprised if you find all students migrating to one corner before the presentation of arguments even begins!)

Key points. Begin this strategy by discussing the item in question. After it is clear that students understand what is intended by the item prompt, call on individuals or groups to suggest key points that should be included in a complete answer. Write the key points on the board as phrases or individual words that will scaffold students' revision, rather than complete sentences they might mindlessly copy. When students return to their responses, have them underline each of the key points they originally included in their answers, then add anything they missed.

Revision with color. Another way that students can revise their answers after a key-points discussion is to use colored pens or pencils and the three Cs. As they read over their responses, they *confirm* correct information by underlining with a green pen; they *complete* their responses by adding information that was missing using a blue pen; and they *correct* wrong information using a red pen.

Review and critique anonymous student work. Use examples of student work from another class, or fabricate student work samples that emulate the problems students in your class are having. Project the work, using an overhead projector, a document camera, or an interactive whiteboard. Have students discuss the strengths and weaknesses of the responses. This is a good strategy to use when first getting students to write in their notebooks. It helps them understand expectations about what and how much to write.

Line of learning. You can have students use a line of learning to show how their thinking has changed. When students return to the original work (embedded or benchmark) to revise their understanding of a concept, they start by drawing and dating a line of learning under the original writing. The line of learning delineates students' original, individual thinking from their thinking after a class or group discussion has helped them reconsider and revise their thoughts.

Find the problems. Review the codes you have recorded and let students know how many items they got right and how many need to be corrected. Give them time in class to work in groups or individually to find their mistakes and correct them. They can turn in the assessment again or write a few sentences in their notebooks explaining how their thinking has changed.

Next-Step Strategies

Group consensus/whiteboards. Have students in each group (or pairs in each group) work together to compare their answers on selected I-Check questions or key points of a notebook entry. First they create a response that the group agrees is the best answer. Groups write their responses on a whiteboard. When you give a signal, one student in each group holds up his or her whiteboard and compares answers. The class discusses any discrepant answers.

Class debate. A student volunteers an answer to an item on an assessment (usually one that many students are having trouble with or one that elicits a persistent misconception). That student is in charge of the debate. He or she puts forth an answer or explanation. Other students agree or disagree, and must provide evidence to back up their thinking. Students are allowed to disagree with themselves if they hear an argument during the discussion that leads them to change their thinking. You can ask questions to keep the discussion on track, but otherwise you should stay on the sidelines.

Critical competitor. Use the critical-competitor strategy when you want students to attend to a specific detail. You present students with two things that are similar in all but one or two aspects. You can use any medium: two drawings, two pieces of writing, or a combination (such as a diagram compared to a description). The point is to compare two pieces of communication or representations in some way that will help students focus on an important detail they might be missing.

Sentence frames. After completing other self-assessment activities, have students consider all the items on the assessment and write a short reflection using sentence frames. This strategy directs students to choose one or two items that they would like to tell you more about.

I used to think _______, but now I think _______.

I should have gotten this one right, but I just _______.

I know _______, but I'm still not sure about _______.

The most important thing to remember about _______ is _______.

Can you help me with _______?

I made a careless mistake, because I know _______.

I'm still confused about _______.

Next time, I will remember to _______.

Now I know _______.

Strategies for Individual Students and Small Groups

Feedback notes. As you read through students' notebooks, add self-stick notes with comments or questions that help guide students to further reflect on and improve their understanding.

Response log. Set up a response log at the back of students' science notebooks (before or after the index or glossary if those are used). Fold a notebook page in half, or draw a line down the center of the page. Have students write "Teacher Feedback" at the top of the left side of the page and "My Responses" at the top of the right side of the page. When you want a student to think about something in his or her notebook, write your note in the "Teacher Feedback" column (or students can move a self-stick note from another page to the response log). Students then respond in the right column, either addressing your comment there or telling you which page to turn to in order to see how they have responded.

NOTE

You can also create feedback notes for embedded assessments using FOSSmap.

Conferences. Use silent-reading time or other times when students work independently to confer with small groups or individual students.

Centers. Set up a center at which students can continue to explore their ideas and refine their thinking. You might pair students so that a student who understands the concept well works with another student who needs some help.

Reteach or clarify a concept. Set up a modified investigation in which a small group of students works with the concept again.

RESPONSE LOG

Teacher Feedback	My Responses
Inv. 1, Part 1 Notebook Entry—You did a good job describing the different sizes of rock you found. Did you find anything else that might have once been living?	Oh, yes, I did, but I thought that was a mistake! Now I know that soil is made up of rock and once living things called humus.
Response Sheet Inv. 1—Great explanation about why same rocks at both locations and why they are smaller downstream. Now how do you explain more rounded rocks downstream?	It's the same reason as why there are smaller rocks. The more they bang together the more rounded they become. The sharp edges get worn off.

FOSSmap

and Online Assessment

FOSSmap is the assessment management program designed specifically for teachers using the FOSS Program in grades 3–8. This system allows you to open online assessments for students, to review codes for student responses, and to run reports to help you assess student learning. FOSSmap was developed at the Lawrence Hall of Science in conjunction with the Berkeley Evaluation and Assessment Research (BEAR) Center at the University of California, as part of a 5-year research and development project funded by the National Science Foundation. It is based on the tools developed in the ASK project.

To use FOSSmap, you must log in to FOSSweb on ThinkLink.

In ThinkLink, assessments are not available to students until you assign them. You will assign them in the same way you make other assignments. The assessments will always be available to you within the course to preview at any time. Responses are sent to the FOSSmap teacher program where the majority are automatically coded. You will need to code short answers and open-response items. Students can answer open-response items on the computer or use paper and pencil, depending on how extensive their typing skills are. There is a drawing tool in FOSSmap, so students can draw models and diagrams online. All items can be answered in the FOSSmap system; none must be done on paper.

For more detailed information on how to use FOSSmap, visit the Knowledge Base on ThinkLink.

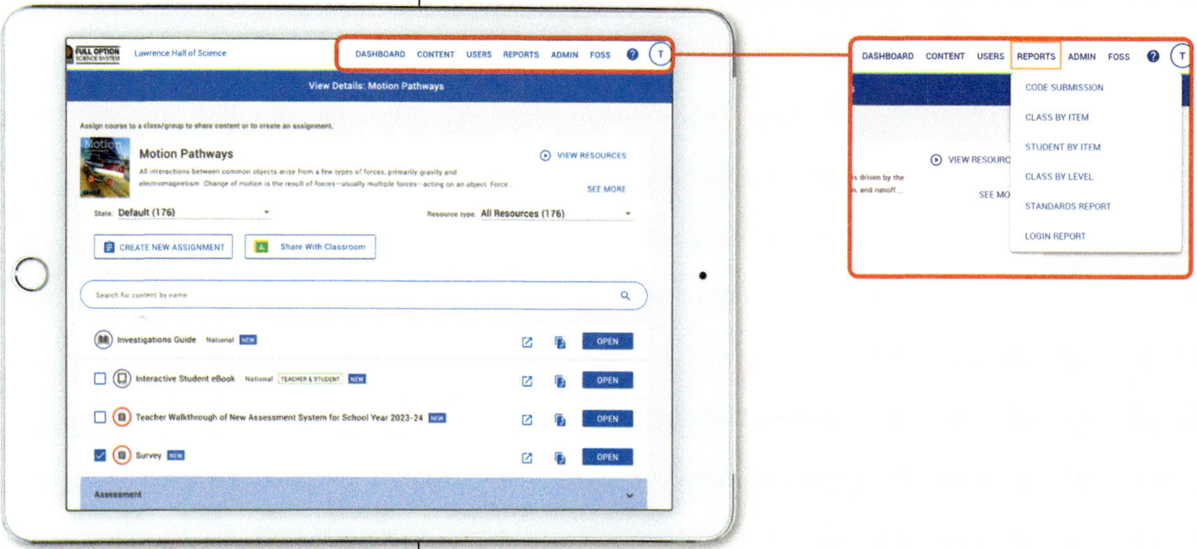

FOSSmap Home Page

FOSSmap Reports

The **Class by Item Report** shows the detail of each item and students' responses. You can go directly to the problem items (indicated by the **Code Frequency Summary Chart**) to get more information and plan next steps.

The **Code Frequency Summary Chart** is a snapshot of how well the students in the class did on the items. Any item where the majority of students did not achieve recognition indicates that the item may have been difficult for students. Recognition and above are considered to be performing at grade level or higher.

The **Student by Item Report** is a good report to share with parents. It shows the student's overall level as well as how students performed on each item. It shows the code means ("What you know and can do to improve") to both teachers and students.

The **Class by Level Report** shows a snapshot of overall class performance on the assessment, with the number of students indicated at each mastery level. Recognition and above are considered to be performing at grade level or higher. Below the graph are the students' names at each level.

The **Standards Report** shows the class performance on each performance expectation assessed in the selected assessment(s). The percentages show the total of students' codes divided by the highest possible codes. It also shows individual student performance on each performance expectation. This can help identify standards that are problematic for each student.

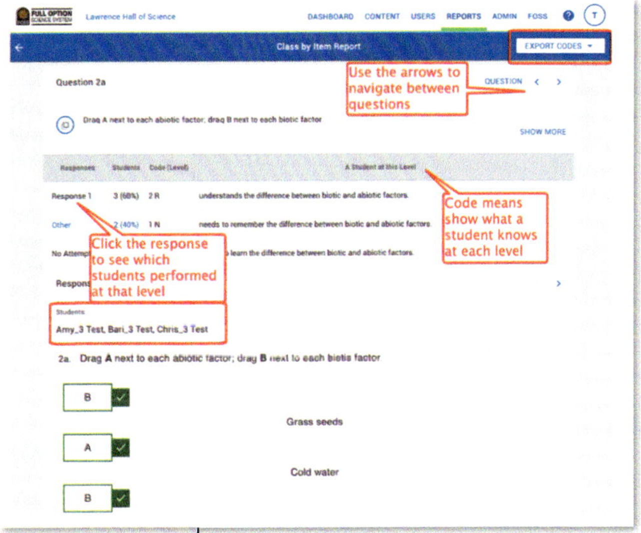

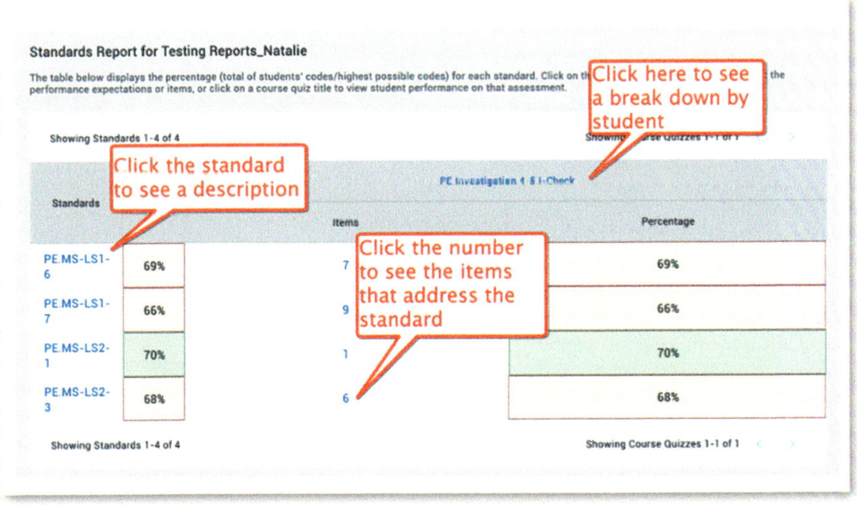

The Standards Report displays progress on NGSS performance expectations.

Sample Assessment Items

FOSS PATHWAYS: SOILS, ROCKS, AND LANDFORMS MODULE

Survey/Posttest

ANSWERS

Name

6. Scientists found the fossil remains of a crocodile in a layer beneath the surface of the Sahara desert where it is very hot and dry. In the same layer they found fish fossils. In a deeper layer further down they found fossils of large trees and now extinct birds.

Which organism lived the longest ago in what is now the Sahara desert?

- Trees and extinct birds
- Fish and crocodiles

Write a claim about how the environment a long time ago was different from what is now the Sahara. Support your claim with evidence.

Key Points

- The environment was different a long time ago.
- The illustration shows there is not much rain in the desert; if this were always true, crocodiles and trees could not have lived there.
- The crocodile fossil provides evidence that this area was once under water or very swampy.
- The trees indicate that at another time, it wasn't as wet, but still wetter than the current environment.

FOSS PATHWAYS

SOILS, ROCKS, AND LANDFORMS MODULE
Survey/Posttest
Page 6 of 8

Item 6

This item provides evidence that students can construct an explanation about how patterns of rock formations and fossils in rock layers indicate changes in the landscape over time. (Contributes to 4-ESS1-1)	
Code (Level)	If the student . . .
4	writes that the environment changed; provides evidence: • the picture indicates little rain in the desert, so not much can grow • the crocodile in one layer shows that at one time the environment was under water • the tree in another layer shows that at another time, there was rain and soil for large trees to grow
3	writes that the environment changed, as in code 4; does not include supporting detail or includes minor errors.
2	writes that the environment changed; provides little or no supporting evidence.
1	writes anything else related to the topic.
0	makes no attempt or writes something off topic.

Key:

- Focus on Science and Engineering Practices
- Focus on Disciplinary Core Ideas
- Focus on Crosscutting Concepts

Codes and corresponding levels

4 - Strategic
3 - Conceptual
2 - Recognition
1 - Notions

NOTE

Look for these resources in the Assessment section in Teacher Resources on FOSSweb.

- Assessment chapter
- Assessment masters
- Coding guides
- Assessment charts

NGSS

Performance Expectations

"The NGSS are standards or goals, that reflect what a student should know and be able to do; they do not dictate the manner or methods by which the standards are taught. . . . Curriculum and assessment must be developed in a way that builds students' knowledge and ability toward the PEs [performance expectations]" (*Next Generation Science Standards,* 2013, page xiv). The FOSS assessment system includes embedded, performance, and benchmark assessments. This chart provides an overview of these assessments across the three fourth-grade modules. These assessments meet the goals of the NGSS by helping students build knowledge and ability in conjunction with active investigations and readings.

Grade 4 NGSS Performance Expectations	FOSS Module	
	Embedded Assessment	Benchmark Assessment
4-PS3-1. Use evidence to construct an explanation relating the speed of an object to the energy of that object.	**Energy** • Inv 3, Part 1: notebook entry • Inv 3, Part 2: response sheet	**Energy** • Investigations 2–3 I-Check • Survey/Posttest
4-PS3-2. Make observations to provide evidence that energy can be transferred from place to place by sound, light, heat, and electric currents.	**Energy** • Inv 1, Part 1: notebook entry • Inv 1, Part 2: notebook entry • Inv 1, Part 3: performance assessment • Inv 2, Part 1: response sheet • Inv 2, Part 2: notebook entry • Inv 2, Part 3: performance assessment	**Energy** • Investigation 1 I-Check • Investigations 2–3 I-Check • Survey/Posttest
4-PS3-3. Ask questions and predict outcomes about the changes in energy that occur when objects collide.	**Energy** • Inv 3, Part 2: response sheet	**Energy** • Investigations 2–3 I-Check • Survey/Posttest
4-PS3-4. Apply scientific ideas to design, test, and refine a device that converts energy from one form to another.	**Energy** • Inv 1, Part 2: notebook entry • Inv 1, Part 3: performance assessment • Inv 2, Part 1: response sheet • Inv 2, Part 2: notebook entry	**Energy** • Investigation 1 I-Check • Investigations 2–3 I-Check • Survey/Posttest
4-PS4-1. Develop a model of waves to describe patterns in terms of amplitude and wavelength and that waves can cause objects to move.	**Energy** • Inv 4, Part 2: performance assessment	**Energy** • Survey/Posttest
4-PS4-2. Develop a model to describe that light reflecting from objects and entering the eye allows objects to be seen.	**Energy** • Inv 4, Part 1: response sheet	**Energy** • Survey/Posttest
4-PS4-3. Generate and compare multiple solutions that use patterns to transfer information.	**Energy** • Inv 2, Part 2: notebook entry • Inv 2, Part 3: performance assessment	**Energy** • Investigations 2–3 I-Check • Survey/Posttest

Grade 4 NGSS Performance Expectations	FOSS Module	
	Embedded Assessment	Benchmark Assessment
4-LS1-1. Construct an argument that plants and animals have internal and external structures that function to support survival, growth, behavior, and reproduction.	**Senses and Survival** • Inv 1, Part 1: notebook entry • Inv 3, Part 1: notebook entry • Inv 2, Part 2: notebook entry • Inv 3, Part 2: response sheet	**Senses and Survival** • Investigation 1 I-Check • Survey/Posttest
4-LS1-2. Use a model to describe that animals receive different types of information through their senses, process the information in their brain, and respond to the information in different ways.	**Senses and Survival** • Inv 1, Part 1: notebook entry • Inv 1, Part 2: notebook entry • Inv 1, Part 3: performance assessment	**Senses and Survival** • Investigation 2 I-Check • Survey/Posttest
4-ESS1-1. Identify evidence from patterns in rock formations and fossils in rock layers to support an explanation for changes in a landscape over time.	**Soils, Rocks, and Landforms** • Inv 4, Part 2: notebook entry	**Soils, Rocks, and Landforms** • Investigation 2 I-Check • Survey/Posttest
4-ESS2-1. Make observations and/or measurements to provide evidence of the effects of weathering or the rate of erosion by water, ice, wind, or vegetation.	**Soils, Rocks, and Landforms** • Inv 1, Part 1: notebook entry • Inv 1, Part 2: response sheet • Inv 2, Part 1: notebook entry • Inv 2, Part 2: response sheet • Inv 2, Part 3: performance assessment	**Soils, Rocks, and Landforms** • Investigations 1–2 I-Check • Survey/Posttest
4-ESS2-2. Analyze and interpret data from maps to describe patterns of Earth's features.	**Soils, Rocks, and Landforms** • Inv 3, Part 1: response sheet • Inv 3, Part 2: performance assessment • Inv 4, Part 1: notebook entry	**Soils, Rocks, and Landforms** • Survey/Posttest
4-ESS3-1. Obtain and combine information to describe that energy and fuels are derived from natural resources and their uses affect the environment.	**Energy** • Inv 1, Part 2: notebook entry	**Energy** • Survey/Posttest
4-ESS3-2. Generate and compare multiple solutions to reduce the impacts of natural Earth processes on humans.	**Soils, Rocks, and Landforms** • Inv 4, Part 1: notebook entry • Inv 2, Part 3: performance assessment • Inv 3, Part 2: performance assessment	**Soils, Rocks, and Landforms** • Survey/Posttest
3-5-ETS1-1. Define a simple design problem reflecting a need or a want that includes specified criteria for success and constraints on materials, time, or cost.	**Energy** • Inv 1, Part 3: performance assessment • Inv 2, Part 1: response sheet **Senses and Survival** • Inv 3, Part 1: notebook entry	**Energy** • Survey/Posttest
3-5-ETS1-2. Generate and compare multiple possible solutions to a problem based on how well each is likely to meet the criteria and constraints of the problem.	**Energy** • Inv 2, Part 2: notebook entry	**Energy** • Survey/Posttest **Soils, Rocks, and Landforms** • Investigations 1–2 I-Check
3-5-ETS1-3. Plan and carry out fair tests in which variables are controlled and failure points are considered to identify aspects of a model or prototype that can be improved.	**Energy** • Inv 2, Part 2: notebook entry	**Senses and Survival** • Survey/Posttest